Emerging Trends in
Wireless Communication

Emerging Trends in Wireless Communication

Dr. Ram Krishan

CWP
Central West Publishing

This edition has been published by Central West Publishing, Australia

© 2021 Central West Publishing

For more information about the books published by Central West Publishing, please visit https://centralwestpublishing.com

Disclaimer
Every effort has been made by the publisher and author while preparing this book, however, no warranties are made regarding the accuracy and completeness of the content. The publisher and author disclaim without any limitation all warranties as well as any implied warranties about sales, along with fitness of the content for a particular purpose. Citation of any website and other information sources does not mean any endorsement from the publisher and authors. For ascertaining the suitability of the contents contained herein for a particular lab or commercial use, consultation with the subject expert is needed. In addition, while using the information and methods contained herein, the practitioners and researchers need to be mindful for their own safety, along with the safety of others, including the professional parties and premises for whom they have professional responsibility. To the fullest extent of law, the publisher and author are not liable in all circumstances (special, incidental, and consequential) for any injury and/or damage to persons and property, along with any potential loss of profit and other commercial damages due to the use of any methods, products, guidelines, procedures contained in the material herein.

A catalogue record for this book is available from the National Library of Australia

ISBN (print): 978-1-922617-22-4

About the Editor

Dr. Ram Krishan, presently working as Assistant Professor and Head, Department of Computer Science, Mata Sundri University Girls College, Mansa, Punjab (A Constituent College of Punjabi University, Patiala), India. Dr. Ram obtained his Ph.D. in Computer Science and Engineering from Guru Kashi University, Talwandi Sabo, India in 2017 and M.Tech. in Computer Engineering from Punjabi University, Patiala, India in 2009. He has authored two academic books and has published more than 40 research papers in various international/national journals and conference proceedings, along with book chapters. Dr. Ram has also edited three research books in the field of wireless communication. His research areas include wireless communication, cloud computing and antenna design.

Contents

Preface

Preface

This is a matter of great pleasure to present the edited book titled Emerging Trends in Wireless Communication. As the title of this book implies, its purpose is to utilize a perspective derived from wireless communication to consider the potential uses of emerging technologies in communications. This book comprises eight chapters contributed by academicians, researchers and students. The major focus of this book is on the emerging trends in wireless communication such as 5G wireless communications, IoT, wireless sensor networks, machine learning and antenna design.

I am thankful to my family and friends for supporting the publication of this edited book. Finally, I am thankful to all who have contributed and spared their valuable time for this book.

Dr. Ram Krishan
Assistant Professor and Head,
Department of Computer Science,
Mata Sundri University Girls College, Mansa, Punjab, INDIA.

CHAPTER 1

NON-ORTHOGONAL MULTIPLE ACCESS FOR 5G WIRELESS COMMUNICATION

Sonia and Silki
JCDM College of Engineering, Sirsa, Haryana, India

1.1 INTRODUCTION

In 4G wireless communication networks, orthogonal multiple access techniques (OMA) were developed and used to fulfil the demand of high data transfer rate, low latency and high bandwidth. However, today's requirements of various services and data, demanding applications are many times higher than that of the 4G era leading to the need for the development of the next generation of wireless communications networks, i.e., 5G. The fifth-generation wireless communication network is an emerging mix of a variety of communication services such as Machine-to-Machine (M2T) communication, Vehicle-to-Everything (V2X), Device-to-Device (D2D), Internet of Things (IoT) and many more. Various technologies will support these communication services in 5G, including millimeter wave radio spectrum, beamforming, massive MIMO and non-orthogonal multiple access (NOMA). 5G is based upon millimeter wave (mmWave) radio spectrum that lies between 24 GHz- 100 GHz. It is a new and less used band. The high-frequency wave carries much more data than the lower frequency wave. The millimetre-wave range makes it possible to have massive connectivity, high data transfer rates, improved capacity using massive MIMO technology. Beamforming makes the transmission between the users and base station (BS) more directional and provides high accuracy, good coverage and low interference of data to be transmitted in the downlink and uplink processes of the wireless mobile communication.

In recent years, NOMA has become a popular choice for researchers as it provides high spectral efficiency, ultra-low latency and massive connectivity as compared with the existing orthogonal multiple access technique OFDMA. There are two main types of NOMA as Power domain NOMA and Code domain NOMA. In the Power domain, NOMA multiplexing is done with different power levels, and in the

Code domain, NOMA multiplexing is carried out with different codes [1].

In this chapter, the concept of the Power domain non-orthogonal multiple access (NOMA) method for the upcoming 5G networks has been explored.

1.2 NEED OF NOMA

Next-generation 5G wireless communication networks and beyond will require massive connectivity, high spectral efficiency, fast data transfer speed and seamless connectivity to support advanced multimedia applications, such as ultra-high-definition video, virtual reality and Wi-Fi integration. As a result, mobile traffic will increase rapidly, and the allotted radio spectrum will become more crowded.

The solution lies in NOMA which provides non-orthogonal access to users in terms of time, frequency and code. Thus, users can access the allotted radio spectrum at the same time, frequency and code but at a different power level so that requirement of advanced properties of next-generation wireless communication networks can be achieved effectively.

NOMA can be integrated with other 5G technologies such as mmWave, small cell, beamforming, massive MIMO to fulfil the need of advanced services like Internet of Things (IoT), cooperative communications, network coding, space-time coding, Vehicle-to-Everything (V2E) communication, Machine-to-Machine (M2M) communications.

1.2.1 NOMA

The key idea of NOMA is to use different power levels for the users to access the base station (BS). The level of power assigned to each user is based on the channel conditions as the distance from the base station, gain, signal to noise ratio (SNR) and fading. Accordingly, near users will be assigned a low-power level, and far users will be assigned a high-power level to maintain user fairness.

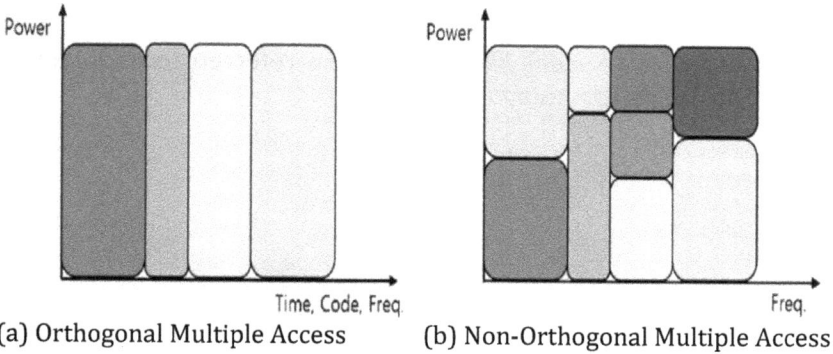

(a) Orthogonal Multiple Access   (b) Non-Orthogonal Multiple Access

Figure 1.1 Difference between OMA and NOMA techniques. Reproduced from [2] with permission.

Figure 1.1 shows the concept of orthogonal multiple access (OMA) and non-orthogonal multiple access (NOMA). The OMA technique provides orthogonal access to the user in terms of time, frequency and code, whereas NOMA is free from the condition of orthogonality. It provides simultaneous access to all subcarriers on the same frequency but at different power levels. Hence, NOMA performs power domain multiplexing of users. The basic processes involved in NOMA include superposition coding (SC) at the transmitter and successive interference cancellation (SIC) performed at the receiver.

1.2.2 NOMA in Downlink

In the downlink process, the signal is transmitted from the base station (BS) to users. Figure 1.2 presents the downlink system of NOMA with 2 users and one base station (BS). The base station transmits the data for both users using superposition coding (SC). In superposition coding, both user's signal is added after multiplying with their corresponding power allocation factor. The near user having a superior channel quality due to less distance between the user and BS will be assigned low power allocation coefficient. The far user having poor channel quality due to large distance of the user from BS will be allotted a high power allocation factor to promote fairness.

Let $x1$ and $x2$ are the signals transmitted from base station to user 1 and user 2 respectively. The power allocation coefficient for user 1 is $\alpha1$ and for user 2 is $\alpha2$. The valuex of $\alpha1$ and $\alpha2$ can be selected arbitrarily such that $\alpha1 + \alpha2 = 1$ and $\alpha2 > \alpha1$.

$h1$ is the channel gain for user 1 (near user) and $h2$ is the channel gain for user 2 (far user). These are often referred to as Rayleigh fading coefficients assuming $|h1|^2 > |h2^2|$.

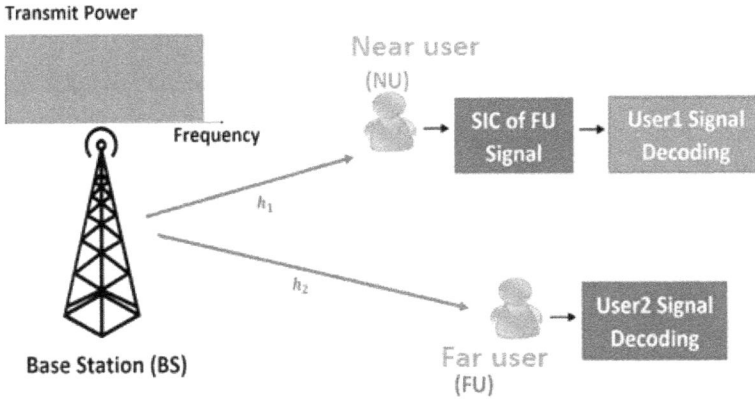

Figure 1.2 Downlink set up in NOMA. Reproduced from [3] with permission.

Let X is superposition coded NOMA signal that is transmitted by BS and is given by equation 1:

$$X = \sqrt{P}\left(\sqrt{\alpha 1}x1 + \sqrt{\alpha 2}x2\right) \tag{1}$$

where, P denotes the transmitted power. Near user will receive signal $y1$ after propagation through channel 1 and is given by equation 2:

$$y1 = h1X + w1 \tag{2}$$

Far user will receive signal $y2$ after propagation through channel 2 and is given by equation 3:

$$y2 = h2X + w2 \tag{3}$$

Here, $w1$ and $w2$ represent the Gaussian noise (AWGN) interference in channel 1 and channel 2 respectively.

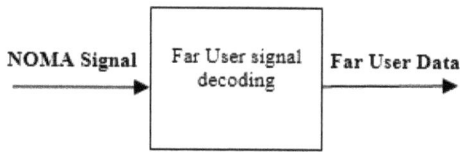

Figure 1.3 Far user processing.

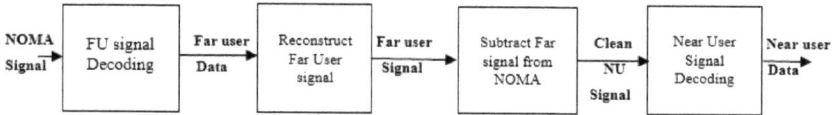

Figure 1.4 Near user processing.

Now the receiver at the far user will decode the received signal $y2$ considering the near user's signals as interference which can be neglected as it is too weak to be decoded. The process is shown in Figure 1.3. On the other hand, the receiver at the near user will receive the composite superposition coded signal (NOMA signal) which is the combination of signals intended for the far user as well as the near user. It will perform the surface interference cancellation (SIC) process to detect its signal (near user). The complete process for near user data processing is shown in Figure 1.4. First of all, the signal intended for the far user will be decoded and regenerated. It will be the stronger signal between the two signals due to the high power allocation factor. This regenerated far user signal is subtracted from the composite superposition coded signal (NOMA signal) resulting in a near user signal. This signal is then decoded to detect the original near user signal. If there is perfect SIC at the near user then it will decode its data with high accuracy as far user interference is completely cancelled in the decoding process.

1.2.3 NOMA in Uplink

In the uplink process, all users transmit their data to the base station. Figure 1.5 presents the uplink NOMA process with 2 users and one base station. User 1 and user 2 transmit their signals to the base station in the same time and frequency slots.

The base station will receive the signal Y given by equation 4:

$$Y = \sqrt{P1}h1x1 + \sqrt{P2}h2x2 + w \tag{4}$$

Here, $P1$ and $P2$ are the independent and non-identical transmitted powers of users. $h1$ and $h2$ are the Rayleigh fading channel coefficients (channel gains). $x1$ and $x2$ are the modulated signals of user 1 and user 2 respectively. w is the Gaussian noise (AWGN) interference.

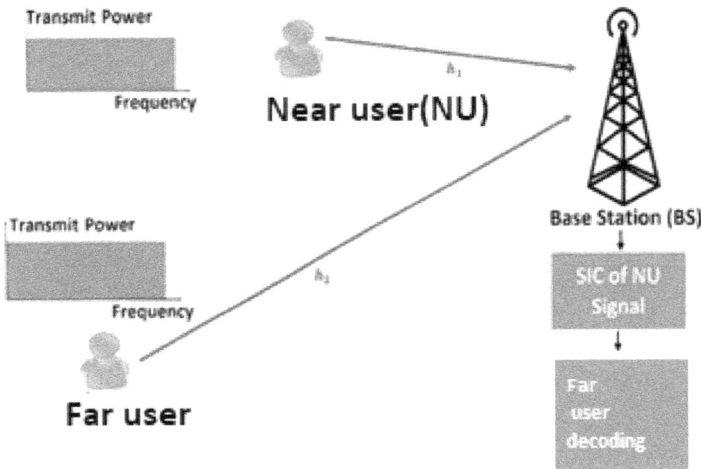

Figure 1.5 Uplink set up in NOMA. Reproduced from [3] with permission.

In the uplink process, the base station first decodes the signal of user 1 considering user 2's signal as interference. Subsequently, by using successive interference cancellation (SIC), data of user 1 is subtracted from the total received signal Y.

Now, from the remaining signal after the SIC process, signal of user 2 is decoded. In the uplink NOMA process, the signal decoded for user 2 (weak/far user) is a clean signal with zero interference, but the signal of user 1 (strong/ near user) suffers from interference from user 2's signal, whereas in downlink NOMA, user 1 (strong/ near user) receives clean signal with zero interference and user 2 (weak/far user) suffers from the interference of user 1's signal.

1.2.4 Benefits of NOMA

In recent years, non-orthogonal multiple access techniques have gained more attention over the orthogonal multiple access in several aspects, such as:

- **Highly spectrum efficient:** NOMA serves multiple users at the same time and with the same frequency but with different power levels, thus, yielding high spectral efficiency as compared to the OMA technique.
- **Massive connectivity:** The number of users is increased under the NOMA scheme because users can access the radio resources at the same time and frequency.
- **High channel capacity:** According to Shannon Hartley channel capacity theorem, NOMA has a high channel capacity than OMA, as in NOMA each user shares the whole allotted bandwidth.
- **Lower latency:** In NOMA, there is a simultaneous transmission of data for each user in the frequency and time domains, so the transmitted signal will experience low latency.
- **User fairness:** In NOMA, user fairness is maintained between the intra-cell user (near user) and cell-edge user (far user), thus, yielding a high throughput for the cell-edge user and enhanced overall performance.

1.2.5 Limitations of NOMA

Apart from various advantages of non-orthogonal multiple access schemes, NOMA suffers from many disadvantages like:

- **Receiver complexity and significant energy consumption:** In a non-orthogonal multiple access scheme, before decoding its signal, each user has to decode the signal from all other users having poor channel gain. This process leads to additional receiver circuitry at the receiver side, hence, a large amount of energy is also consumed in this process.
- **High inter-user interference:** Non-orthogonal multiple access techniques allow high inter-user interference as the same resources are allotted to each user at the same time and frequency.

- **The limited effective number of users:** If there is an imperfect SIC process for one user, then it will lead to erroneous decoding of all other users, thus, limiting the users within a cell, which will reduce the sum-rate gain of the NOMA scheme.
- **Sensitivity towards uncertainty in channel gain measurement:** In NOMA, each user has to deliver its channel gain information back to the base station, and the NOMA scheme is sensitive towards uncertainty in the measurement of channel gain.

1.3 RELATED WORK

In the present era, increasing demand for the high data rate, high capacity, low latency and massive connectivity will lead the wireless communication networks towards the next generation (5G and beyond) [4,5]. Various techniques in 5G like mmWave [6], MIMO [7,8], beamforming [9] and NOMA [10] have emerged and can be combined to meet the high spectral efficiency, high energy efficiency and ultra-low latency requirements of the different communication networks.

Next-generation communication services as the Internet of Things (IoT), Vehicle-to-Everything (V2X), Machine-to-Machine (M2M) and advanced multimedia applications, such as ultra-high-definition video and virtual reality, will lead towards enormous data traffic in a cellular network [11]. To fulfil the requirements of these next-generation wireless communication networks, NOMA technique is proposed which can be combined with the other 5G communication technologies.

Orthogonal multiple access technique serves users on different radio resources like time and frequency, whereas non-orthogonal multiple access provides access to the users on the same time, frequency and codes but with different power level multiplexing. In the NOMA scheme, superposition coding is applied at the transmitter side, and successive interference cancellation is applied at the receiver side both in the downlink and uplink process. Mostly, it is assumed that SIC is error-free [12].

In recent years, NOMA has been studied in terms of power allocation [13,14], system capacity and capacity comparison between OMA and NOMA using Shannon Hartley capacity theorem [15], outage probability [16] and bit error rate (BER). The BER performance of NOMA both in the uplink and downlink processes has been demonstrated with different modulation schemes as well as different fading channels [17,18].

Further, the performance of NOMA has been studied when it is integrated with various wireless communication techniques such as multiple-input multiple-output (MIMO), cooperative relaying, simultaneous wireless information and power transfer (SWIPT) [19,20].

In this chapter, NOMA has been simulated using the MATLAB platform for a better understanding of its basic process, followed by its performance evaluation with different parameters.

1.4 SIMULATION SET-UP

Table 1.1 provides the simulation setup used in this chapter to evaluate the performance of NOMA.

Table 1.1 Simulation set up

Parameter	Values
Power allocation schemes	Fix
Power allocation Coefficients	$\alpha 1 = 0.75, \alpha 2 = 0.25$
Modulation scheme	BPSK, QPSK, Combination of BPSK and QPSK
Distance of users from BS	Far User= 1000 m; Near User= 500 m
Power	0 to 40 dBm
Path loss exponent (eta)	eta = 4
Noise	AWGN with zero mean value
Fading channel	Rayleigh Fading Channel
No. of Monte Carlo Simulations (N)	N=5*10^5
Target Rates for both users	Far User= 1, Near User =2

1.4.1 Simulation Results

As discussed in section 1.2.1, NOMA is accomplished in three steps:

o   Super position-coding at BS transmitter.
o   SIC at near user and detection of near user data.
o   Detection of far and near user data.

In this section, the simulation results of these three steps have been presented. It is assumed that data transmitted from BS for far (user 1) and near (user 2) users are: $x1 = [1010]$; $x2 = [0110]$.

The power allocation coefficients for far and near users are $\alpha 1 =$ 0.75; $\alpha 2 =$0.25. The modulation scheme used for signal transmission to both users is BPSK. The superposition coded data after scaling with power allocation coefficients is shown in Figure 1.6.

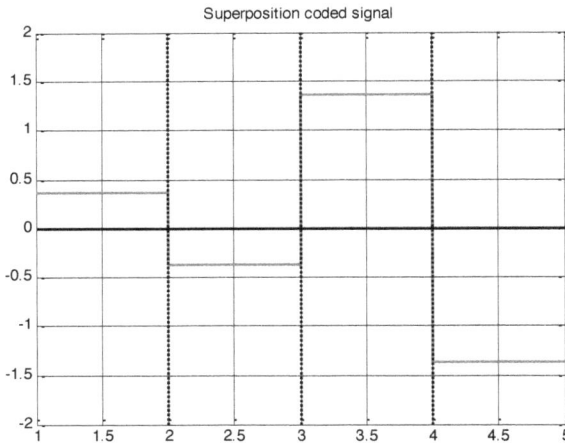

Figure 1.6 Superposition coded signal.

Figure 1.6 Superposition coded signal.

This signal is received by both users. Now, SIC is performed at the near user, and the remaining data $(Xrem)$ is shown in Figure 1.7.

SIC Performed at near user/remaining data after SIC

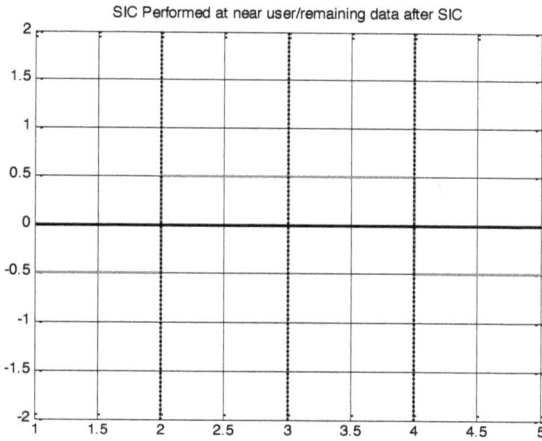

Figure 1.7 Signal after performing SIC.

Figure 1.8 provides the data at far user and near user after detection. It is observed that the data received by both users are the same as transmitted from BS.

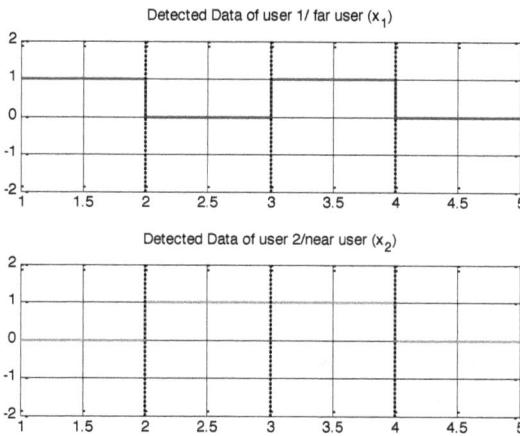

Detected Data of user 1/ far user (x_1)

Detected Data of user 2/near user (x_2)

Figure 1.8 Detected data for far and near user.

1.4.2 Performance Evaluation

In this section, the performance evaluation of NOMA is presented for the variation in modulation scheme, power allocation factor and distance from BS. The performance parameters considered are capacity, achievable rate, BER and outage probability.

I. Variation in modulation schemes: In this section, the performance of NOMA is evaluated for different modulation schemes like BPSK, QPSK and a combination of both. The modulation scheme employed at both users may be the same or different. So, in this section, the performance of NOMA is evaluated for the three different cases:

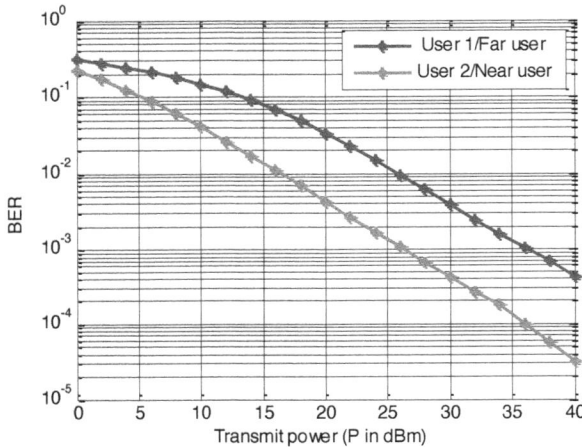

Figure 1.9 BER for BPSK modulation for both users.

a. Both users employing BPSK: The modulation scheme used by the near user and far user is BPSK. Figure 1.9. shows the comparison of BER achieved at the near user and far user. It is observed that the BER of near user is lower than that of far user.

b. Both users employing QPSK: Figure 1.10 shows the comparison of the bit error rate of both users employing QPSK as a modulation scheme. It is indicated that the BER of near user is lower than that of far user.

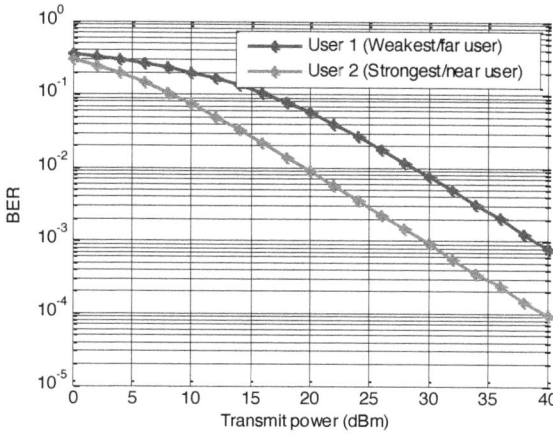

Figure 1.10 BER for QPSK modulation for both users.

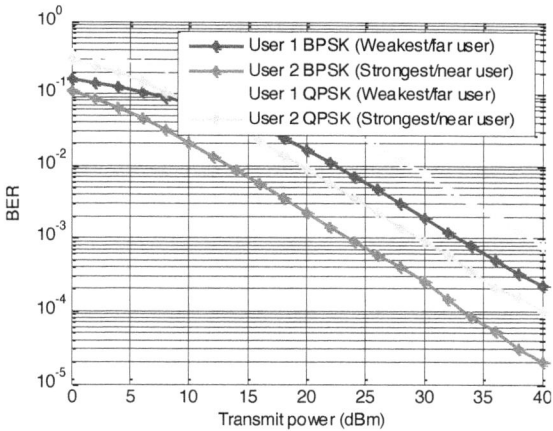

Figure 1.11 BER comparison (BPSK vs QPSK).

Figure 1.11 shows the comparison of BER for both users using BPSK and QPSK modulation schemes. It can be gathered that BPSK offers a lower BER than the QPSK modulation scheme for both users.

c. BPSK at far user and QPSK at near user: This section provides the simulation results when different modulation schemes are employed for far user and near user. The modulation scheme assigned to the far user is BPSK whereas the near user is assigned QPSK for modulation.

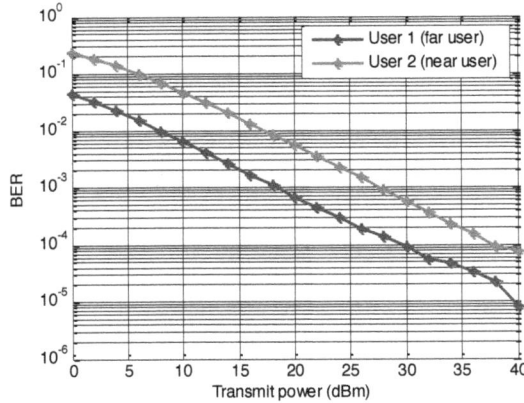

Figure 1.12 BER for far user (BPSK) and near user (QPSK).

Figure 1.12 shows the simulation results for BER of near and far users for the case. It is observed that the BER of the far user is lower than that of near user in this case, which indicates that a high modulation level should be selected for good channel conditions, whereas a low modulation level should be selected for the poor channel conditions to sustain a reliable BER performance. In NOMA, far users experience more severe channel conditions than near users, so BPSK modulation can be employed for the far users, while QPSK modulation for the near users can be a better choice.

II. Variations in power allocation factors: In this section, the performance of NOMA has been evaluated for different values of power allocation factors, i.e., $\alpha 1$ and $\alpha 2$, with a fixed distance of both users from BS, i.e., d1= 1000, and d2=500 for far user and near user, respectively. The performance parameters considered for comparison are outage probability, capacity, achievable rates and bit error rate. Three scenarios have been considered to evaluate the performance of NOMA.

a. Scenario 1: In this scenario, four performance parameters have been evaluated at $\alpha 1 = 0.75, \alpha 2 = 0.25$

Table 1.2 shows the comparison of average outage probability, capacity, achievable rate and bit error rate over 25 observations for both users. It is observed that average capacity and average achievable rates of near user are higher than the far user, while average outage probability and average BER of near user are better than the far user.

Table 1.2 Comparison of outage probability, capacity, achievable rate and bit error rate in scenario 1

User	Average outage probability	Average Capacity (Mbps)	Average achievable rate (bps/Hz)	Average BER
Far user	0.50	1.101	1.101	0.157
Near user	0.48	6.663	6.663	0.110

b. Scenario 2: The same simulation set-up is run again for 25 observations at $\alpha 1 = 0.8$, $\alpha 2 = 0.2$.

Table 1.3 shows the comparison of average outage probability, capacity, achievable rate and bit error rate for both users. The simulation results show the superior performance parameters as compared to scenario 1.

Table 1.3 Comparison of outage probability, capacity, achievable rate and bit error rate in scenario 2

User	Average outage probability	Average Capacity (Mbps)	Average achievable rate (bps/Hz)	Average BER
Far user	0.50	1.27	1.27	0.153
Near user	0.49	6.45	6.45	0.112

c. Scenario 3: In this scenario, four performance parameters have been evaluated at $\alpha 1 = 0.5, \alpha 2 = 0.5$.

Both users have been allocated the same power coefficients. Table 1.4 shows the comparison of average outage probability, capacity, achievable rate and bit error rate over 25 observations for both users. It is observed that $\alpha 1 = 0.5$, $\alpha 2 = 0.5$ is the worst-case scenario, as shown in Table 1.4. It is observed that both users will be in an outage at every value of given power. Capacity, achievable rates and BER of both users are worse than in scenarios 1 and 2. From the above discussion, it is concluded that the far user should be allotted a high power allocation factor whereas the near user should be allotted a low power allocation factor.

Table 1.4 Comparison of outage probability, capacity, achievable rate and bit error rate in scenario 3

User	Average outage probability	Average Capacity (Mbps)	Average achievable rate (bps/Hz)	Average BER
Far user	1	0.57	0.57	0.30
Near user	1	7.40	7.40	0.27

III. Variations in distance of BS from both users: In this section, the performance of NOMA has been evaluated for variation in distance of near user and far user from the base station. The performance is evaluated with respect to the bit error rate, achievable rate, capacity and outage probability in three different scenarios as large, moderate and equal distance from the base station. The power allocation factors are fixed in all the scenarios as $\alpha1$ =0.75, $\alpha2$ =0.25

a. Large distance: In this scenario, the distance of the far user from the base station is considered much greater as compared to the near user (far user=1000 m, near user=500 m). Observation for this case is same as Table 1.2. From Table 1.5, it is observed that the achievable rates and capacity of near user are higher than far user, whereas the outage probability and bit error rate of near user is better than that of far user.

Table 1.5 Performance evaluation for large distance

User	Average outage probability	Average Capacity (Mbps)	Average achievable rate (bps/Hz)	Average BER
Far user	0.50	1.101	1.101	0.157
Near user	0.48	6.663	6.663	0.110

b. Moderate distance: In this scenario, the distance of the far user from the base station is considered much greater as compared to the near user (far user=500 m, near user=300 m). Table 1.6 shows the performance of NOMA implemented at both users under different performance parameters. It is indicated that the performance of

NOMA is improved significantly when the distance of both users from BS is decreased. This is due to the reason that decreasing the distance between the users and BS improves the quality of the propagation channel between them.

Table 1.6 Performance evaluation for moderate distance

User	Average outage probability	Average Capacity (Mbps)	Average achievable rate (bps/Hz)	Average BER
Far user	0.196	1.538	1.538	0.0589
Near user	0.160	9.015	9.015	0.0255

c. Equal distance: In this scenario, the distance of both users from the base station is considered equal (far user=500 m, near user=500 m). Table 1.7 shows the performance of NOMA implemented at both users for different performance parameters. It is indicated that the performance of the far user is the same as that in scenario 2, but for near user, it is decreased as the near user is allotted a low power allocation factor which results in its degraded performance as compared to the far user. Thus, it can be concluded that the performance of the users depends upon the power allocation factor as well as their distance from BS.

Table 1.7 Performance evaluation for equal distance

User	Average outage probability	Average Capacity (Mbps)	Average achievable rate (bps/Hz)	Average BER
Far user	0.196	1.54	1.54	0.0589
Near user	0.475	6.67	6.67	0.110

1.5 CAPACITY COMPARISON BETWEEN NOMA AND OMA

As mentioned in previous sections, a NOMA scheme is a better candidate as compared to traditional OMA in terms of capacity. In this section, the performance of both systems has been compared based on capacity which is of prime concern. According to the Shannon channel capacity theorem, the capacity of a communication channel is given by equation 5:

$$C = Wlog_2(1 + SNR)bits/sec \tag{5}$$

The Shannon capacity gives the achievable rates in a certain channel, with certain bandwidth (W) and a certain signal to noise ratio (SNR). In OMA, each user takes half the bandwidth. On the other hand, in NOMA, all users share the complete bandwidth of the channel but with different power levels. Figure 1.13 shows the basic idea of OMA and NOMA.

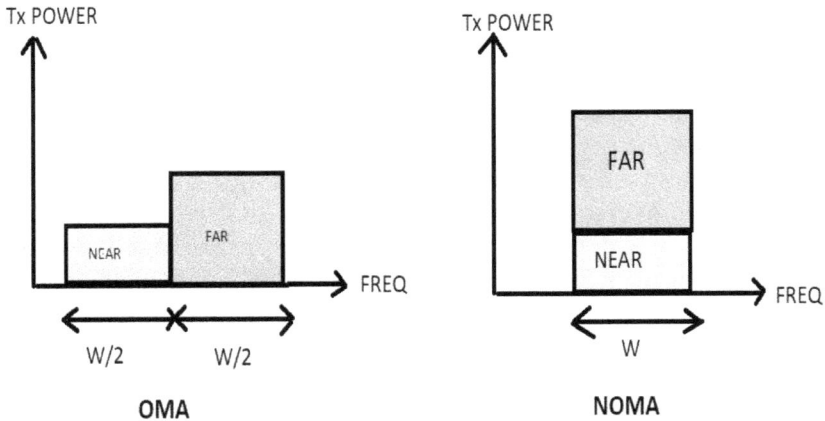

Figure 1.13 OMA vs. NOMA.

Now signal to noise ratio (SNR) for both schemes is given by:

SNRs for OMA: $SNR_{near} = \frac{P_{near}}{Noise}$ and $SNR_{far} = \frac{P_{far}}{Noise}$

SNRs for NOMA: $SNR_{near} = \frac{P_{near}}{Noise}$ and $SNR_{far} = \frac{P_{far}}{P_{near}+Noise}$

Thus, the expression of capacity for both schemes will become:

$$C_{OMA} = \frac{W}{2}log_2\left(1 + \frac{P_{near}}{noise}\right) + \frac{W}{2}log_2\left(1 + \frac{P_{far}}{noise}\right)bits/sec \tag{6}$$

$$C_{NOMA} = w \, log_2 \left(1 + \frac{P_{near}}{noise}\right) + w \, log_2 \left(1 + \frac{P_{far}}{P_{near}+noise}\right) bits/sec$$

$$(7)$$

$$P_{far} + P_{near} = P_{total} = constant \qquad (8)$$

Total capacity for the OMA scheme is given by equation 6 and for NOMA scheme is given by equation 7. As shown in equation 8, the total power for the near and far users is always constant.

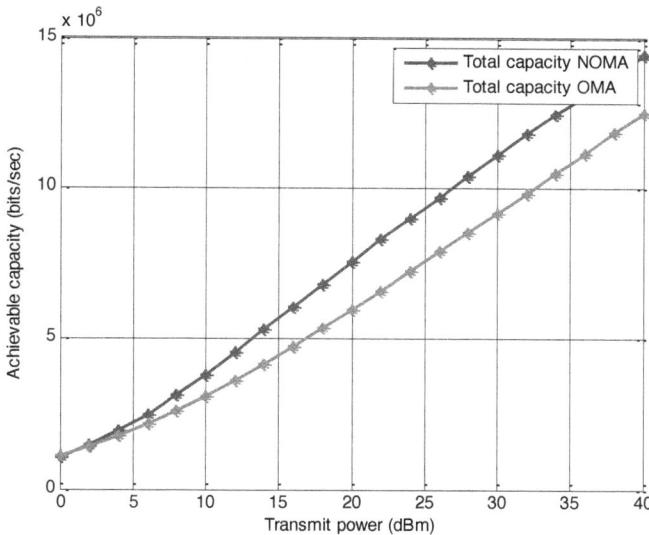

Figure 1.14 The capacity comparison (NOMA vs. OMA).

Figure 1.14 provides the simulation results of the comparison of capacity in OMA and NOMA with respect to input power. It is observed that the capacity of OMA is lesser as compared to NOMA, as also mentioned in the literature.

1.6 CONCLUSION

NOMA is an important enabling technology for achieving the 5G key performance requirements, including high channel capacity, high system throughput, ultra-low latency and massive connectivity. This chapter presented the basic concepts of NOMA including superposition coding at transmitter site and surface interference cancellation at receiver side both in the uplink and downlink processes. The impact of various design factors including modulation schemes, distance from the base station and power allocation factor has been studied by simulation. The simulation is carried for downlink NOMA. From the simulation results, it is observed that BPSK yields lesser BER for far and near users than the QPSK modulation scheme. Further, it is observed that BER can be improved when QPSK is applied to the near user and BPSK is applied to the far user. In addition, performance analysis of NOMA for other performance parameters as outage probability, achievable rates and capacity is also performed using simulations. From simulation results, it is observed that the outage probability, achievable rates, capacity and BER of near user are superior than that of far user. In the end, capacity comparison between NOMA and OMA has been demonstrated.

References

1. https://futurenetworks.ieee.org/tech-focus/june-2017/noma-in-5g-systems.
2. Cheon, J., and Cho, H. (2017) Power allocation scheme for non-orthogonal multiple access in underwater *Acoustic Communications Sensors*, **17**(2465), 1-13.
3. Kara, F., and Kaya, H. (2018) BER performances of downlink and uplink NOMA in the presence of SIC errors over fading channels. *IET Communications*, **12**(15), 1-11.
4. Boccardi, F., Heath, R. W., Lozano, A., Marzetta, T. L., and Popovski, P. (2014) Five disruptive technology directions for 5G. *IEEE Communications Magazine*, **52**(2), 74–80.
5. Andrews, J. G., Buzzi S., Choi, W., Hanly, S. V., Lozano, A., Soong, A. C. K., and Zhang, J. C. (2014) What will 5G Be? *IEEE Journal on Selected Areas in Communications*, **32**(6), 1065–1082.
6. Agrawal, S. K., and Sharma, K. (2016) 5G millimeter wave (mmWave) communications. *2016 3rd International Conference on Computing for Sustainable Global Development (INDIACom)*, India, pp. 3630-3634.
7. Ding, Z., Schober, R., and Poor, H. V. (2016) A general MIMO framework for NOMA downlink and uplink transmission based on signal alignment. *IEEE Transactions on Wireless Communications*, **15**(6), 4438–4454.
8. Choi, J. (2016) On the power allocation for MIMO-NOMA systems with layered transmissions. *IEEE Transactions on Wireless Communications*, **15**(5), 3226–3237.
9. Golbon-Haghighi, and Mohammad-Hossein (2016) Beamforming in wireless networks. In: *Towards 5G Wireless Networks - A Physi-*

cal Layer Perspective, Bizaki, Hossein Khaleghi (ed.), IntechOpen Limited, UK, pp. 163-192.

10. Kizilirmak, R. C. (2016) Non-orthogonal multiple access (NOMA) for 5G networks. In: *Towards 5G Wireless Networks - A Physical Layer Perspective*, Bizaki, Hossein Khaleghi (ed.), IntechOpen Limited, UK, pp. 83-98.

11. Multiple Liu, Y., Qin, Z., Elkashlan, M., Ding, Z., Nallanathan, A., and Hanzo, L. (2017) Nonorthogonal access for 5G and beyond. *Proceedings of the IEEE*, **105**(12), 2347–2381.

12. Islam, S. M. R., Avazov, N., Dobre, O. A., and Kwak, K. (2017) Power-domain non-orthogonal multiple access (NOMA) in 5G systems: Potentials and challenges. *IEEE Communications Surveys & Tutorials*, **19**(2), 721–742.

13. Oviedo, J. A., and Sadjadpour, H. R. (2017) A fair power allocation approach to NOMA in multiuser SISO systems. *IEEE Transactions on Vehicular Technology*, **66**(9), 7974–7985.

14. Cui, J., Liu, Y., Ding, Z., Fan, P., and Nallanathan, A. (2018) Optimal user scheduling and power allocation for millimeter wave NOMA systems. *IEEE Transactions on Wireless Communications*, **17**(3), 1502–1517.

15. Aldababsa, M., Toka, M., Gökçeli, S., Kurt, G. K., and Kucur, O. (2018) A tutorial on nonorthogonal multiple access for 5G and beyond. *Wireless Communications and Mobile Computing*, **2018**, 1–24.

16. Wang, J., Xia, B., Xiao, K., Gao, Y., and Ma, S. (2018) Outage performance analysis for wireless non-orthogonal multiple access systems. *IEEE Access*, **6**, 3611–3618.

17. Jain M., Soni S., Sharma N., and Rawal D. (2019). Performance analysis at near and far users of a NOMA system over fading channels. *2019 IEEE 16th India Council International Conference (INDICON)*, India, pp. 1-4.

18. El-Mokadem, E. S., El-Kassas, A. M., Elgarf, T. A., and El-Hennawy, H. (2019) BER performance evaluation for the downlink NOMA system over different fading channels with different modulation schemes. *2019 5th International Conference on Science and Technology (ICST)*, Indonesia, pp. 1-6.

19. Makki, B., Chitti, K., Behravan, A., and Alouini, M. S. (2020) A survey of NOMA: Current status and open research challenges. *IEEE Open Journal of the Communications Society*, **1**, 179–189.

20. Thirunavukkarasu, M., Sparjan, R., and Thangavelu, L. (2020) An adaptive power allocation scheme for performance improvement of cooperative SWIPTNOMA wireless networks. *Computers, Materials & Continua*, **63**(2), 1043–1064.

CHAPTER 2

INTERNET OF THINGS (IoT) – FUTURE ASPECTS AND CHALLENGES

Jaspal Singh
Department of Physics, Mata Sundri University Girls College, Mansa-151505, Punjab, India

In the past decade, people have witnessed exponential growth in the field of technology. There is no field which has not been impacted by this aggressive advancement. Internet of Things (IoT) is one of the most talked-about technologies in this era. A system of interconnected and intelligent devices with or without human interaction was just a dream a few decades ago. However, in the past few years, with the advent and advancement of self-updating algorithms, this idea has been incubated into reality. Starting with a program developed from scratch, IoT has reformed health systems, automation in the industry, agriculture practices and almost everything else. Moreover, IoT is expected to have limitless potential through machine learning (ML), networking agility, automation, security reforms and web-scale markets. However, the future is still obscure. This chapter aims to cover the basic structure of an IoT architecture with respect to the futuristic aspects and vulnerabilities of this technology. Nowadays, most of the IoT systems are designed in a hierarchal way. The algorithms based on data from scratch are governed by a set of superior learning methodologies in a horizontal manner. Therefore, interoperability becomes quite complex. Such architectures not only need high computational hardware but also add up to the cost of the system. Connecting numerous machines, which are based on artificial intelligence (AI), could provide a solution to such barriers. Internet of Senses is another progression that can be added. On the parallel side, security is going to be the prime challenge to IoT systems. By 2025, it is expected that there will be a significant population (approximately 21 billion) of IoT devices. Hence, there will be plenty of risk from the cybercriminals. Botnet-powered distributed denial of service (DDoS) attacks will also get smarter. Detailing all these advancements and challenges/concerns is the main focus of this chapter.

2.1 INTRODUCTION

Internet of Things (IoT) is not an outlandish term these days. Definition of this system says that any network consisting of interconnected devices, computers, digital machines, animals, etc., which have unique identities, is known as an IoT system. In the field of technology, the network of interconnected entities has opened new opportunities for researchers. In this era of know-how, the information technology sector has shown enormous growth. With the advent of the 5G communication networks and wireless technology, the future is going to witness an outstanding feat of technology. In the coming decade, the IoT technology will provide a backbone to futuristic information technologies and communications. In addition, the wireless sensor networks (WSNs) are already in the existence and are going to be the hot topic for researchers. Numerous WSN based IoT applications are there which have attracted technocrats. Whenever IoT is discussed, the very basic idea which comes to mind is the architecture of the network. It not only provides the blueprint of the technology but also decides numerous features of IoT. As in an IoT architecture, WSNs are the main building blocks which are also known as the sensing units or nodes. Being of a discrete nature, such entities raise hopes as well as concerns about the future of information technology and communication. Depending upon the capability of these sensory nodes, one can decide about the capability, reach and limits of a network. Moreover, various computational constraints can also be treated as an inter-related part of such architecture. The indispensable applications of such WSNs include the term *Smart* with the name of the conventional applications. The concept of smart cities, smart transportation and traffic systems, smart agriculture, etc., is nothing but applications of such entities and IoT networks. Thus, to estimate the future of such systems, it becomes the need to understand how this system works. Figure 2.1 shows a generalized IoT-based system.

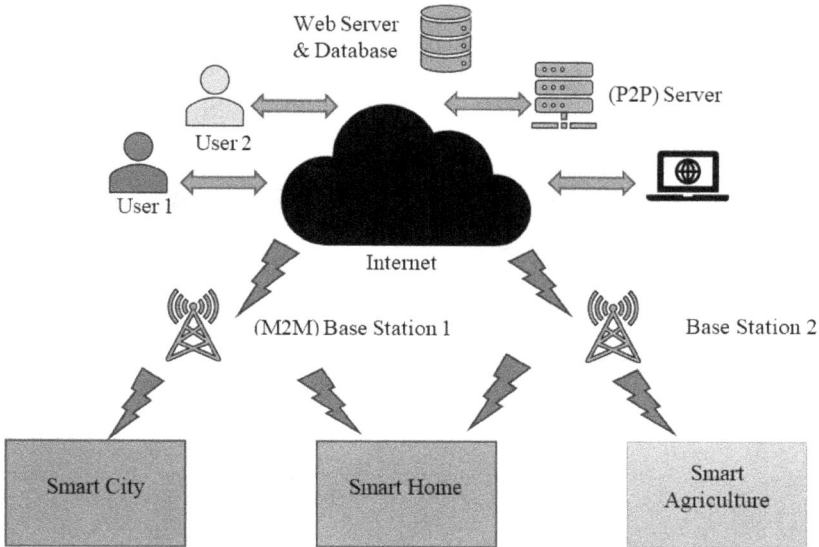

Figure 2.1 Generalized architecture of a IoT based system.

Figure 2.1 depicts that WSNs are the basic building blocks of the IoT systems. Figure 2.2 shows the layered architecture of a WSN [1]. It also depicts that how the various network layers are clustered in a WSN. Five layers namely the application layer, transport layer, network layer, data link layer and physical layer are present in this architecture which handle all the responsibilities related to data transfer. Moreover, three cross layers are also present to handle the power management, mobility management and task management, respectively. The details of the functions performed by these layers are out of the scope of this paper. However, a detailed explanation can be found in [1]. Other than this architecture, one can also refer to the Cluster Network Architecture. However, the explanation of the same is also beyond the scope of this chapter.

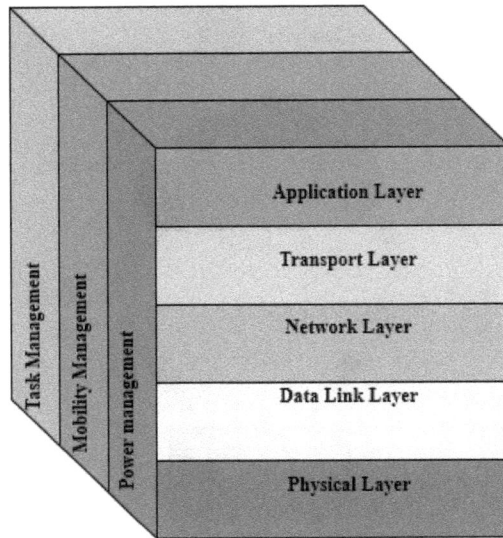

Figure 2.2 Layered architecture of a wireless sensor network.

2.2 HISTORY OF IoT

The abbreviation IoT was first introduced by Ashton K. [2] and his pears in 2009. This group of researchers is the founder group of the original MIT Auto-ID Laboratory. Just after one year, LG announced its first internet-connected refrigerator in 2000. Thereafter in 2005, the UN's International Telecommunication Union (ITU) called IoTs a new dimension in the field of ICT. Later in 2015, the term Internet of Anything has been introduced which has impacted almost all sectors [3]. Ashton K. mentioned almost a decade ago that the whole digital world was dependent on humans for information exchange and data generation which was around 50 petabytes. However, over the past years, technocrats have even explored the Internet of Nano-Things (IoNT) which has opened an entry for this technique into daily lives by incorporating the internet into the personal, professional and societal life of mankind. Moreover, the addition of a virtual world of inanimate quasi-intelligent devices is also expected to augment the future [4].

2.3 IoT ECOSYSTEM

The IoT technological ecosystem consists of several microprocessors like Arm Cortex-M, Arc, Quark, etc. These are the electronic components that are governed by various operating systems like (uCLinux, Embedded Linux, Android Auto, Ubuntu, TinyOS, etc.). Moreover, this ecosystem consists of numerous collectors-cum-aggregation devices which consist of network routers, access nodes, ZETA platforms, etc. [5], which are integrated using different infrastructures like Cisco lox (fog) [6], BeagleBone [7] and minicomputers (like raspberry and Arduino). Two indispensable components of this ecosystem are interoperability and IoT protocols. Interoperability defines various rules and regulations related to open interconnection aiming at the development of advanced technologies and platforms for the certification of various nodes/devices which are used to form a IoT system. An example of such a device is IOBridge. Second term represents the protocol which governs the communication in a IoT system. These integrating frameworks used for different functions in a network are defined as the protocols. Some of the prime and important protocols are Bluetooth, Cellular, DDS, DSRC, HTTP, Ethernet, Z-Wave, NFC, SATCOM, etc. [5].

2.4 CATEGORIZATION OF IoT

As already established, there is no limit in the scope of IoT. Several applications are nowadays incorporating IoT. Based on these applications, Internet of Things can be classified as:

- Industrial Internet of Things (IIoT)
- Consumer Internet of Things (CIoT) (also called Human IoT)
- Internet of Everything (IoE)

This classification is adopted based on Moor Insights and Strategy Report [8] which is just a broad classification. If the categorization is done based on the application, there would be more than 1600 IoT-based independent projects, as claimed by Scully [9].

Table 2.1 IoT based projects

IoT Segment	Global Share of IoT Projects	Details		
		Americans	European	APAC
Smart City	23%	34	45	18
Connected Industry	17%	43	31	20
Connected Building	12%	53	33	13
Connected Car	11%	54	30	12
Smart Energy	10%	42	35	19
Other	8%	50	34	11
Connected Health	6%	55	29	15
Smart Supply Chain	5%	49	36	12
Smart Agriculture	4%	39	26	31
Smart Retail	4%	53	35	9

IoE is used by many organizations. Cisco and Qualcomm are the prime names in this regard [10], [11]. The term IoE is considered to be based on the four basic entities which are data, people, processes and things. On the contrary, IoT is considered to be based on only 'things'. Figure 2.3 explains the term IoE and how various entities are interlinked, as given by D. Evans [12].

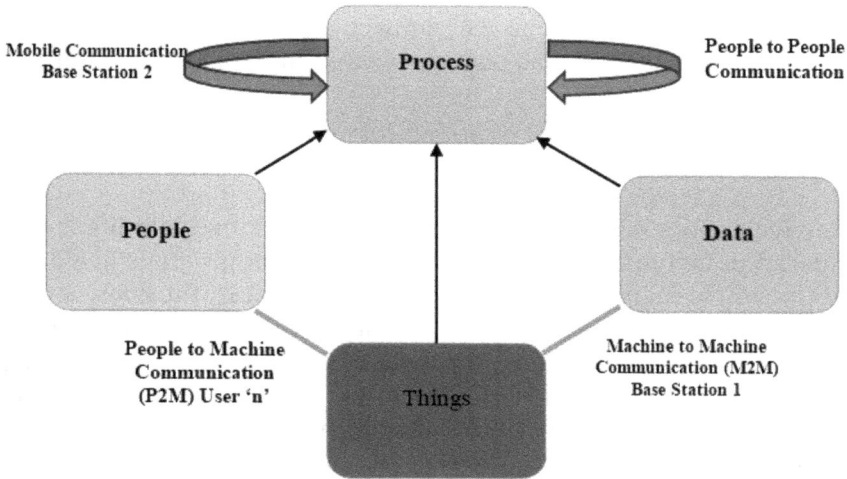

Figure 2.3 Internet of Everything: A generalized architecture.

2.5 FUTURE APPLICATIONS OF IoT

As established already, IoT is not a separate entity from the internet. Instead, it is the expansion of the same. To understand the system hierarchy, let us consider an application when someone has to attend a meeting at 9 AM. Let us consider that the person has an IoT-based system at home and living in a smart city as well. In case due to the unknown reasons the meeting gets preponed or postponed by a certain time, the smartwatch will allow adjusting the wake-up alarm accordingly as everything will be connected. Thereafter, the whole day schedule may automatically be preponed or postponed as well. At home, the person will be notified about the traffic situation at the updated time and accordingly route to the office will be updated for the person. In another scenario, the smart wearable sensors will let the person know if he is feeling well or needs some medical attention. Thus, everything will be convenient as it would not require any manual update. This is the main concept of IoT. The backbone for the same is the internet.

In this context, the cellular communication can be regarded as the heart of the IoT systems. With the introduction of the 5G cellular networks, it emerged as a technology that provides a limitless connectivity for the IoT demands, thus, providing reliable and secure coverage all over the globe. The impact on the various fields owing to such advancements has been presented in detail in the later sections.

2.5.1 Smart Farming

Any technology which does not have an impact on daily life is considered to be futile. As agriculture is a field that provides livelihood to the whole mankind, IoT would not be worth it if it does not impact this sector. Owing to this, the agriculture sector is going to witness a drastic advancement. As nowadays the farmers are struggling to cope up with the yield loss due to various factors like malnutrition, pest infestation, soil nutritional imbalance, rainfall irregularities, etc., thus, precision agriculture is inevitable. According to the data published in various reports at IoT World Congress, the size of the agricultural economy is expected to increase up to $15.3 Billion by the year 2025, owing to the introduction of Smart Agriculture [13]. IoT solutions will be more concentrated on expanding the awareness among the farmer community and introducing them to the supply chain. This will be ensured that they know about supply and demand gaps. Improving the yield and profitability will be the major focus of the technology. Other than this, preserving the natural ecosystem is another vision. IoT in agriculture will require a complete transformation of the existing equipment and will need specialized instrumentation having wireless connectivity. Figure 2.4 provides a generalized overview of an agricultural IoT system.

Many practices have incorporated IoT for various purposes in the agricultural sector. An android/mac platform-based application 'Plantix' [14] can be named as one of the dominant utilization of IoT. Other studies have also focused on the prediction of the nutrient requirements of the crops [15], [16].

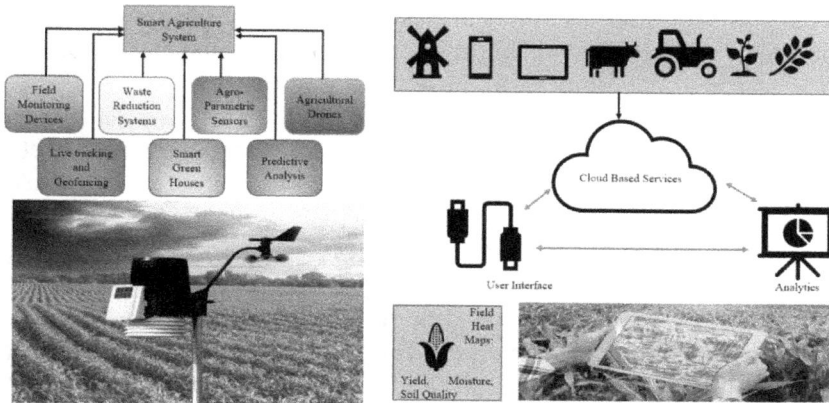

Figure 2.4 A smart agricultural system: weather conditions, GPS (automated steering), crop health and soil quality.

Other than the mentioned utilities, Table 2.2 summarizes the potential applications of IoT in agriculture [17].

Table 2.2 IoT involvement in smart agriculture

Advanced Agriculture	**Monitoring Applications**	Soil Moisture, Soil Health, Crop Health, Crop Diseases, Livestock Population, Machinery, Hydroponics Systems, Vertical Farming, Phenotyping, Smart Diary, Irrigation Control, Milk Farm Monitoring
	IoT based Services	Irrigation, Fertilization, Fungicide/Weedicide/Pesticide/Herbicide Usage, Soil Preparation and Treatment, Crop Yield and Storage maintenance
	Sensors with Different Functionality	Temperature, Plant Leaf and Stem, Fruit Quality and Size, Root, Animal Feed Input, Disease Detection of Crops

2.5.2 Industrial Automation

Modern manufacturing industries are being shifted to Industry 4.0 platforms from the traditional platforms. According to a survey taken by Delloite Global [18], IoT is one of the top-ranked technologies, which has the highest impact on the manufacturing industry. Further, it is concluded in the survey that IoT, artificial intelligence,

Cloud, and big data analytics are the four biggest players with the capability of providing a bedrock for numerous organizations to gather data and work on more intelligent operations.

In this context, cost reduction, revenue growth, security and safety, quality control, etc., are going to be the main drivers of the IoT-based projects in near future. These drivers will offer better decisions leading to increased efficiency, data processing from interconnected systems targeting revenue-boosting, remote monitoring, control of operations and tracking abnormalities. Moreover, focusing on the cyber-physical systems, Industry 4.0 has the capability to acquire and process data. Smart industry (specifically manufacturing industry) is concentrating to implement the following factors for getting Smart [19]:

- Smart products (self-processing, self-explanation and self-interacting based on machine learning)
- Smart maintenance (IoT sensor-based data generation and data-driven maintenance)
- Smart materials (smart labelling and RF identification tags)
- Smart metrics (advanced intelligent measurements and analysis for zero loss productivity)
- Smart manpower (robotic arms, high-efficiency work plans)

2.5.3 IoT Based Revenue Opportunities in Near Future

As soon as the industry realizes the benefits of IoT, it tends to shift from a money-consuming venture to a revenue generator. To generate maximum output, the industry will have to go through the monetization opportunities. These instances will be usually driven by the four models which will consider data, insights, platforms and business as a service. Figure 2.5 summarizes these models which will be governing revenue growth in the coming years using IoT.

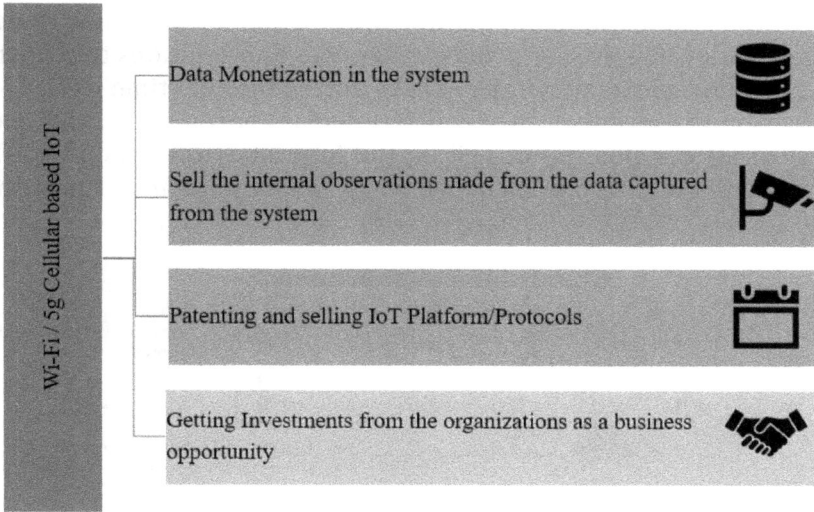

Figure 2.5 IoT based revenue generating models.

2.5.4 Digital Twin Technology

A digital twin is a technology that enables us to create a digital equivalent of an object/system or thing in a virtual world. To date, it is just an introductory concept. However, by the end of 2026, it is expected to grow at a rate of 58% [20]. In this system, several sensors are put on to an object/thing/system which continues to observe the real-time data throughout the lifetime of the entity. These sensors are linked to an IoT system which comprises various technologies like big data processing, cloud computing or artificial intelligence. This system generates a virtual equivalent in the digital domain of a system in the physical world based on real-time data. The real-time data might be in the form of health status, performance parameters or anything which can be accessed via IoT or 4G/5G cellular network sensors. The key benefit of this technology is that it will help scientists to acquire and analyze data which was not available to date. An example is the study of the environmental conditions leading to the extinction of various wild creatures. Figure 2.6 depicts how a digital twin can be obtained from an entity of the physical world.

This system can transform the world. The people will be able to design and develop precise products, services and solutions to numerous problems as they will be able to access the real-time data. The provision of simulation in the digital domain is something that has been practiced before but will be the futuristic application of IoT. The health sector is one such filed which will be benefitted the most.

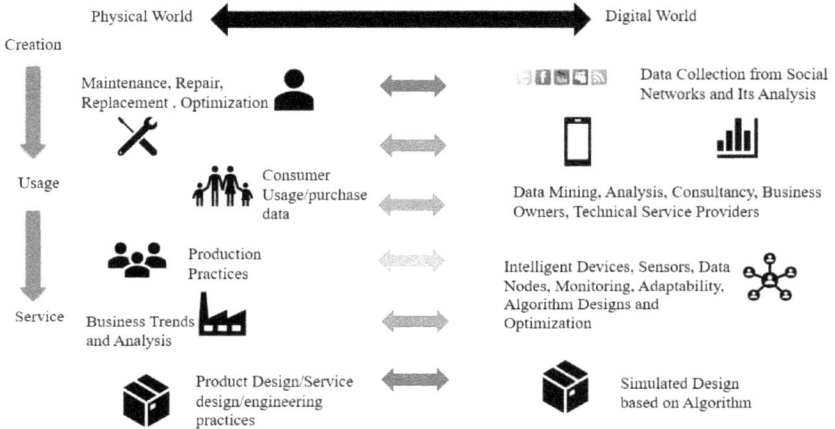

Figure 2.6 Digital twin design process.

2.5.5 Energy and Utility

IoT has pushed the limits further by impacting the energy and utilities being used in today's world. It is expected that people will soon be able to use the energy systems that will be digitally controlled and will allow the full-duplex flow of information and power. These IoT-based systems will allow the users to control and optimize the energy demand. With the invention of low-cost renewable power resources (solar panels), it has become possible to trade the power both ways. A consumer will also be able to sell the power to the retailers with the help of an automated supply generation system and transmission lines. Moreover, digital solutions will also be provided to the retail companies to reduce the input cost, automated bill payment, etc. Also, both-way communication will allow improving stability/reliability as the system will be equipped with high precision sensors which can avoid and handle energy crises. Utilities are being designed keeping IoT, robotics and automation in mind. The future utilities will be focusing on:

- Outage optimization
- Detecting and avoiding theft of power/energy
- Planning to design service packages as per consumer demand
- Improving consumer satisfaction based on the data analysis collected using IoT systems

Table 2.3 gives a detailed description of IoT applications in different energy-related areas.

Table 2.3 IoT application areas in energy sector [21]

Application	Sector	Description
Smart Grid	Electrical	Big data and ICT based technology-based design platform
Network Management	Electric Grid Design and Management	Different operating nodes based on big data
Control	Electric Grid Design and Management	Analysis of load and supply using intelligent devices
Vehicle Grid Control	Electrical	Charge/discharge cycles of electric instruments
Microgrids	Electrical	Independent small grid management than the parent one
Control and Management of Network Heating Issues	Heating Management	Temperature and load analysis
Demand Response Data Collection and Management	Residential and Commercial	Centralized policy development based on big data
IoT based Energy Metering	Consumer Community	Intelligent devices for acquiring data related to load and heating
Emergency Battery Man-	Consumer Community	Battery triggering at optimal conditions

agement		
Smart Infra-structure	Consumer Community	Centralized and optimal control at user and supplier end

Other than this, reversing the natural resource loss (especially drinking water) by implementing IoT-based diagnostics has already been started in the UK [22]. Such practices will be seen to be improving in terms of their efficacy based on IoT as well.

2.5.6 Gesture Control

Gesture control is a field of signal processing that is nowadays gaining popularity. With the invention of numerous wearable electronic gadgets, this technique will be quite promising in handling the complications related to the human body. Such a device is known as "armband", which is equipped with the sensors which can detect human muscle contractions and relaxations. Ultimately, it becomes feasible to judge the muscle movements and generate commands accordingly. The application of such gesture control is switching on the lights with hand movement. When attached with the internet/cellular data, such appliances will provide great opportunities for smart homes.

2.5.7 Light Control

It is another field in which the application of IoT has already marked its presence. However, the automation of the lighting depending upon the number of people gathered still needs attention. Undoubtedly in near future, this will be one of the most dominant applications of IoT.

2.5.8 Smart Glassware

People will be having smart crockery at homes. The smart glass will be able to access that whether the person has enough amount of water or requires some reminders. In turn, it may also intimate another device in the network if something is not well. For example, a doctor appointment may be booked if the water intake is abnormal and if the wearable electronic gadget has confirmed any change in body vitals.

2.5.9 Body Vital Measurements

It will include heartbeat monitoring, pulse oximeter, blood pressure monitoring, etc. Moreover in some of the projects, people are working on the early enough prediction of schizophrenia, panic and epilepsy attacks [23]–[26], so that a person can be alarmed promptly. All this will be the future of the IoT in the medical industry. Other than this, the following are some of the potential applications of IoT which are likely to emerge and transform the health sector:

- Real-time acquisition and availability of patient data
- Improvement in workplace manpower management via sensory smart designed chips and real-time location tracking systems.
- Use of inter-connected vital response measuring devices at the workplace
- IoT connected devices for monitoring the health status of remote patients
- Supply chain management for pharmaceutical purposes

2.5.10 Miscellaneous Fields

In addition, there are many other fields that are going to be the spectators of IoT technology. A few of them can be listed as:

- Transport and logistics (e.g., tire pressure monitoring system)
- Smart eye (Google Dream Project - The Glass)
- Audio Bridge (wireless Hi-Fi system, which can play audio in a required room after sensing the presence of a human-based on heat maps)
- Robotics (machine learning-based technology which can be the future of manufacturing)
- Automotive applications (e.g., driverless cars)
- Smart air (air quality control devices based on the IoT systems)

2.6 CHALLENGES

There is no skepticism that IoT is the future. As already discussed, there will be a complete transformation of every sector with the availability of the 5G cellular networks. However, the challenges will be growing and cannot be ignored. Some of the dominating issues which will seek immediate attention are discussed in the later sections.

2.6.1 Low Power Sensory Units

As the number of IoT devices will be increasing exponentially, there will be a significant need for powering batteries. However, to date, very few sensors are there whose battery lasts for the whole life span. In this context, the biggest challenge for the researchers will be to develop energy-efficient sensory units as well as protocols. New data structuring techniques based on traditional as well as new WSN architectures will be needed for developing long-lasting sensor nodes. Tables 2.4 and 2.5 summarize a few of the existing protocols which can be used to further optimize the life span of sensor units.

Table 2.4 Existing technique-based routing protocols for energy efficient routing protocols

Sr. No.	Reference	Routing Protocol	Energy Efficiency results
1	[27]	LEACH	Improved lifetime than LEACH and EAD protocols
2	[28]	GREEDY Algorithm	Lesser energy consumption and transfer delay
3	[29]	TMHDHAC	Less Energy dissipation and higher data transfer rate
4	[30]	LEACH-HPR	Better for energy saving
5	[31]	LEACH-D	Greater energy efficacy
6	[32]	DFTBC	35% more saving of energy
7	[33]	PRRP	PRRP prolongs the lifetime of the network

Table 2.5 AI techniques-based routing protocols

Sr. No.	Reference	Routing Protocol	Throughput	Energy Efficiency	Delay
1	[34]	ACO-LEACH	N-AVL	AUG	N-AVL
2	[35]	IEEMARP	AUG	AUG	N-AVL
3	[36]	LEACH-VA	N-AVL	AUG	N-AVL
4	[37]	Hybrid Clus-	N-AVL	AUG	N-AVL

		ter and Chain-Based Routing Technique			
5	[38]	ESO-LEACH (AI)	N-AVL	AUG	N-AVL
6	[39]	Improved LEACH (AI)	N-AVL	AUG	N-AVL
7	[40]	EBRP	N-AVL	AUG	N-AVL
8	[41]	SMEER	AUG	AUG	N-AVL
9	[42]	(FEEC-IIR	AUG	AUG	N-AVL
10	[43]	GWO (IoT)	AUG	AUG	N-AVL
11	[44]	FBFPSTERP	N-AVL	AUG	N-AVL
12	[45]	EERP	N-AVL	AUG	N-AVL
13	[46]	FIEPE	N-AVL	AUG	N-AVL
14	[47]	EEEFL	N-AVL	AUG	N-AVL

AUG-Augmented/Increased/Improved, N-AVL- Not Available

2.6.2 Privacy and Security

As the IoT device count keeps on increasing, there are increasing number of chances of data and privacy theft. The prominent reason behind the same is that the cybercriminals are more aware than the consumer community and are more equipped with tools and software, thus, raising IoT data safety concerns. As per the blog published by Norton [48], IoT devices will also provide a platform for the cybercriminals and hence there will be increased extents of thefts, e.g., like DDoS attacks. As mentioned by Norton, in 2016, people came across a malware named "Internet of Things" which was a malicious software capable of infecting data collecting device nodes in a IoT system. The Mirai malware is another software that can access any network device. Such devices start behaving like a botnet which is responsible for distributed denial of service (DDoS) attacks. Such attacks can bring multinational brands to kneel down as their goodwill is compromised by such attacks. Therefore, the next challenge is that the routers have to be more secure and smart so that they can detect the fake botnet (converted devices). Such malware can have the capability to weaponize the IoT devices as well. 5G will further raise the concerns related to such security standards. The only safety tool will be IoT-trained manpower who can counterattack such malware activities and, hence, secure the networks. The main security-related issues and challenges which will need imme-diate attention include:

- Weak authentication and authorization
- Lack of collective and self-sustaining management
- Limited cryptography- Lack of encryption
- Insecurity at the web architecture level
- Untimely updating of security protocols
- Increased usage of default settings at every level of the sys-tem (e.g., usernames and passwords)

To avoid any such obligations, one can follow the security frame-work as suggested by Deloitte, as given in Figure 2.7.

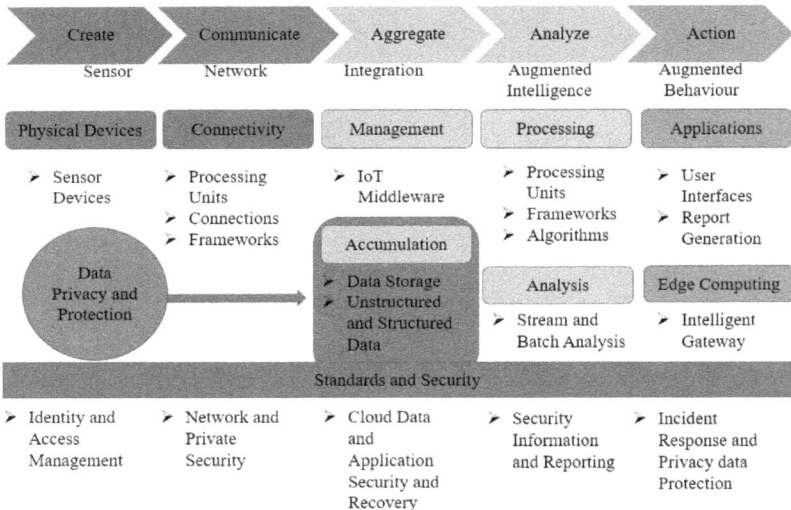

Figure 2.7 Security framework for a IoT based system [18].

2.6.3 Accuracy and Speed

The accuracy of IoT in daily life has been measured in the appropri-ate standards. The involvement of 5G can affect the accuracy as well as speed as a futuristic scope. However, the introduced technologies can have a more qualitative impact on the daily life applications.

2.6.4 Creating and Sustaining Success in Industry

As the techniques evolve, traditional practices can be the biggest challenge. So, for being in the competition and active in the business, being agile and customer-oriented can help a lot. Therefore, creating a successful IoT-based revenue model and maintaining it is one of the significant issues. Such IoT system-based monetization-cum-revenue model can be achieved by:

- Building a digital operating platform (needs a lot of leader-ship commitment)
- Establishing a robust work environment-cum-ecosystem and a quality product that can survive the technological apocalypses.
- Technological stack formation for coping with the technology-related challenges.
- Up to date build-in safety and privacy protocols
  - o One can go for merged perimeter boundaries for privacy
  - o The dynamic nature of the system will help in safety as the nodes in IoT may not be static
  - o Involve more than one stakeholder as the decentralized system can be more secure and safe

2.7 CONCLUSION

After going through this chapter, it becomes clear that the Internet of Things has no limits in its applications and expandability. A vast number of sectors, including agriculture, healthcare, transportation and lodging, fleet management, manufacturing industry, communication, etc., are involved and trying to implement IoT and achieve new milestones. However, the accomplishments are scarce at the moment, and opportunities are expected in the future. Despite all these achievements, many concerns and challenges like privacy, infrastructure, energy demands, etc., are also growing which need immediate and intense attention. Soon, the mankind will be entering a new era of technology where the dreams of today will be a reality.

References

1. https://www.elprocus.com/architecture-of-wireless-sensor-network-and-pplications/
2. Ashton, K. (2009) That "Internet of Things" thing: In the Real World Things Matter More than Ideas. *RFID Journal*, http://www.rfidjournal.com/articles/
3. Barile, M. F. (1983) A brief history of the IOM. *Yale J. Biol. Med.*, **56** (5-6), 353-354.
4. Miraz, M. H., Ali, M., Excell, P. S., and Picking, R. (2018) Internet of nano-things, things and everything: Future growth trends. *Future Internet*, **10**(68), 1-28.
5. Jindal F., Jamar R., and Churi P. (2018) Future and challenges of internet of things. *Int. J. Comput. Sci. Inf. Technol.*, **10**(2), 13–25.
6. Maas, M. (2017) Technical Evangelist, IoT, DevNet DEVNET-2039. Cisco Live Zone, DevNet. https://www.ciscolive.com/c/dam/r/ciscolive/apjc/docs/2017/pdf/DEVNET-2039.pdf.
7. Beagle Bone AI (2020) Online: https://beagleboard.org/bone.
8. M. I. & Strategy (2014) Segmenting the Internet of Things (IoT). Online: https://www.moorinsightsstrategy.com/wp-content/uploads/2014/05/Segmenting-the-Internet-of-Things-IoT-by-Moor-Insights-and-Strategy.pdf.
9. Scully, P. The Top 10 IoT Segments in 2018—Based on 1,600 Real IoT Projects. IoT Anal. Mark. Insights Internet Things. Online: https://iot-analytics.com/top-10-iot-segments-%0A2018-real-iot-projects/#.
10. Weissberger, A. TiECon 2014 Summary-Part 1: Qualcomm Keynote & IoT Track Overview, IEEE ComSoc. Online: https://community.comsoc.org/blogs/alanweissberger/tiecon-2014-summarypart-%0A1-qualcomm-keynote-iot-track-overview.
11. Evans, D. The Internet of Everything: How More Relevant and Valuable Connections Will Change the World Online: https://www.cisco.com/web/about/ac79/docs/innov/IoE.pdf.
12. Evans, D. (2012) How the Internet of Everything Will Change the World. Cisco Blog, 2012. Online: http://blogs.cisco.com/news/how-the-internet-of-everything-will-change-the-worldforthe-%0Abetter-infographic/.
13. IoT Solutions World Congress (2020). Online: https://www.iotsworldcongress.com/iot-transforming-the-future-of-agriculture/
14. Plantix. Online: https://plantix.net/en/
15. Singh, A., and Singh, M. L. (2015) Automated Color Prediction of Paddy Crop Leaf using Image Processing. 2015 IEEE International Conference on Technological Innovations in ICT for Agriculture

and Rural Development. Online: TIAR 2015, 2015, pp. 24–32, doi: 10.1109/TIAR.2015.7358526.

16. Singh, A., and Singh, M. L. (2016) Performance Evaluation of Various Classifiers for Color Prediction of Paddy Plant Leaf. *J. Electron. Imaging*, **25**(6), 061403-10.

17. Khan, N., Ray R. L., Sargani, G. R., Ihtisham, M., Khayyam, M., and Ismail, S. (2021) Current progress and future prospects of agriculture technology: Gateway to sustainable agriculture. *Sustain.*, **13**(9),1–31.

18. Stephen, R., Ayshwarya, B., Shantha Mary Joshitta, R., and Shanthan, H. B. J. (2021) Internet of Things (IoT): The Standard Protocol Suite For Communication Networks. In: Cases on Edge Computing and Analytics. *IGI Global Publisher of Timely Knowledge*, 55–72.

19. Taneja, G., Rishi, R., and Saluja, R. (July 2019) Future of IOT. *PC Quest.* **26**(1), 10–28.

20. Digital Twin. Online: https://www.marketsandmarkets.com/PressReleases/digital-twin.asp.

21. Motlagh, N. H., Mohammadrezaei, M., Hunt, J., and Zakeri, B (2020) Internet of things (IoT) and the energy sector. *Energies*, **13**(2), 1–27.

22. IoT: Powering the Future of Business and Improving Everyday Life (2020). Online: www.cognizant.com/IoT.

23. Singh, K., Singh, S., and Malhotra, J. (2020) Spectral features based convolutional neural network for accurate and prompt identification of schizophrenic patients. *Proc. Mech E, Part H: J of Engineering in Medicine*, 1-18.

24. Singh, K., and Malhotra J. (2018) Stacked Autoencoders based Deep Learning approach for Automatic Epileptic Seizure Detection. 2018 First International Conference on Secure Cyber Computing and Communication (ICSCCC), pp. 249-254. Online: https://www.semanticscholar.org/paper/Stacked-Autoencoders-Based-Deep-Learning-Approach-Singh-Malhotra/7508c05e4000d6880cebe9cb78f001a3b74daeae

25. Singh, K., and Malhotra, J. (2018) IoT enabled Epileptic Seizure Early Detection System using Higher Order Spectral Analysis and C4.5 Decision Tree Classifier.5th International Conference on Computing for Sustainable Global Development (IndiaCom-2018), pp. 1105–1110.

26. Singh. K., and Malhotra J. (2021) Cloud based ensemble machine learning approach for smart detection of epileptic seizures using higher order spectral analysis. *Phys. Eng. Sci. Med.*, (44)1, pp. 313–324.

27. Galzarano, S., Fortino, G., and Liotta, A. (2014) A learning-based MAC for energy efficient wireless sensor networks. Lect. Notes

Comput. Sci. Lect. Notes Artif. Intell. Lect. Notes Bioinformatics, **8729**, 396–406, doi: 10.1007/978-3-319-11692-1_34.

28. Yin, G., Yang, G., Yang, W., Zhang, B., and Jin, W (2008) An energy-efficient routing algorithm for wireless sensor networks. ICICSE 2008 - 2008 International Conference on Internet Computing in Science and Engineering, pp. 181–186. Online: https://www.semanticscholar.org/paper/An-Energy-Efficient - Routing-Algorithm-for-Wireless-Yin-Yang/ b7886cc0c43fbf1d3d728b7cfa43088516bc36a4

29. Jiang, Y., Lung, C. H., and Goel, N. (2010) A Tree-Based Multiple-Hop Clustering Protocol for Wireless Sensor Networks. International Conference on Ad Hoc Networks pp. 371–383, Online: https://link.springer.com/chapter/10.1007/978-3-642-17994-5_25

30. Li, H. (2010) LEACH-HPR: An energy efficient routing algorithm for heterogeneous WSN. *2010 IEEE Int. Conf. Intell. Comput. Intell. Syst. ICIS 2010*, **2**, 507–511. Online: 10.1109/ICICISYS.2010.5658274.

31. Liu, Y., Luo, Z., Xu, K., and Chen, L. (2010) A reliable clustering algorithm base on LEACH protocol in wireless mobile sensor networks. *ICMET 2010 - 2010 International Conference on Mechanical and Electrical Technology*, pp. 692–696, doi: 10.1109/ICMET.2010.5598449.

32. Tan, N. D., and Viet, N. D. (2015) DFTBC: Data Fusion and Tree-Based Clustering Routing Protocol for Energy-Efficient in Wireless Sensor Networks. In: Knowledge and System Engineering, Springer International. Online: https://www.springerprofessional.de/en/dftbc-data-fusion-and-tree-based-clustering-routing-protocol-for/2284608

33. Zaman, N., Tang Jung, L., and Yasin, M. (2016) Enhancing Energy Efficiency of Wireless Sensor Network through the Design of Energy Efficient Routing Protocol. *Journal of Sensors*, **2016**, 1-17.

34. Agarwal, T., Kumar, D., and Prakash, N. R. (2010) Prolonging Network Lifetime Using Ant Colony Optimization Algorithm on LEACH Protocol for Wireless Sensor Networks, Network. *International Conferences on Recent Trends in Networks and Communications*, pp. 634–641.

35. Nayyar, A., and Singh, R. (2020) IEEMARP- a novel energy efficient multipath routing protocol based on ant Colony optimization (ACO) for dynamic sensor networks. *Multimedia Tools and Applications*, **79**, 35221-35252. https://doi.org/10.1007/s11042-019-7627-z

36. Liang, H., Yang, S., Li, L., and Gao, J. (2019) Research on routing optimization of WSNs based on improved LEACH protocol. *EURASIP Journal of Wireless and Communications and Networking*, **194**, 2-12.

37. Hadjila, M., Guyennet, H., and Feham, M. (2014) A hybrid cluster and chain-based routing protocol for lifetime improvement in WSN. WWIC 2014 LNCS 8458, 257–268 Online: 10.1007/978-3-319-13174-0_20.

38. Nigam, G., and Dabas, C. (2018) ESO-LEACH: PSO based energy efficient clustering in LEACH. *Journal of King Saud University - Computer and Information Sciences*, **1**(1), 4–11.

39. Zhang, Y., Li, P., and Mao, L. (2018) Research on Improved Low-Energy Adaptive Clustering Hierarchy Protocol in Wireless Sensor Networks. *J. Shanghai Jiaotong Univ.*, **23**(5), 613–619.

40. Li, L., and Li, D. (2018) An Energy-Balanced Routing Protocol for a Wireless Sensor Network. *Journal of Sensors*, **2018**, Article ID 8505616, 1-13.

41. Dhand, G., and Tyagi, S. S. (2019) SMEER: Secure Multi-tier Energy Efficient Routing Protocol for Hierarchical Wireless Sensor Networks. *Wireless Personal Communications*, **105**(1), 17–35.

42. Sathya Lakshmi Preeth, S. K., Dhanalakshmi, R., Kumar, R., and Mohamed Shakeel, P. (2018) An adaptive fuzzy rule-based energy efficient clustering and immune-inspired routing protocol for WSN-assisted IoT system. *J. Ambient Intell. Humaniz. Comput.*, doi: 10.1007/s12652-018-1154-z.

43. Manshahia, M. S. (2019) Grey Wolf algorithm-based energy-efficient data transmission in internet of things. *Procedia Computer Science*, **160**, 604–609.

44. Mittal, N. (2020) An Energy Efficient Stable Clustering Approach Using Fuzzy Type-2 Bat Flower Pollinator for Wireless Sensor Networks. *Wireless Personal Communications*, **112**(2), 1137–1163.

45. Ghaffari, A. (2014) An Energy Efficient Routing Protocol for Wireless Sensor Networks Using A-Star Algorithm. *Journal of Applied Research and Technology*,**12**(4), 815–822.

46. Nehra, V., Sharma, A. K., and Tripathi, R. K. (2020) FIEPE: Fuzzy Inspired Energy Efficient Protocol for Heterogeneous Wireless Sensor Network. *Wireless Personal Communications*, **110**(4), 1769–1794.

47. Singh, Malik, A., and Kumar, R. (2017) Energy efficient heterogeneous DEEC protocol for enhancing lifetime in WSNs. *Engineering Science and Technology: An International Journal*, **20**(1), 345–353.

48. Norton. The future of IoT: 10 predictions about the Internet of Things. Online: https://us.norton.com/internetsecurity-iot-5-predictions-for-the-future-of-iot.html.

CHAPTER 3

APPLICATIONS OF WIRELESS SENSOR NETWORKS FOR IoT BASED REMOTE HEALTHCARE MONITORING IN COVID-19 PANDEMIC

Ram Singh
Punjabi University Patiala, Punjab, India

3.1 INTRODUCTION

Since January 2020, the outbreak of novel coronavirus disease (COVID-19) has engulfed the entire world. The health services of many countries have been badly affected. Many countries have been struggling to handle the pandemic situation while providing healthcare services to the COVID-19 infected patients. This deadly virus spreads through the human respiratory droplets among the masses via direct contact and with virus-affected surfaces or objects that came into contact with COVID-19 patients. All walks of human life are badly affected. As COVID-19 is a transmittable disease, it becomes very challenging for healthcare workers to safeguard themselves while working in COVID-19 infected wards. In such situations, many doctors, nurses and paramedical workers got affected by the COVID-19 and many of them have lost their lives. As of data available up to April 15, 2020, the medical physicians' deaths in the total deaths due to COVID-19 pandemic have been reported as follows: Italy 44%, Iran 15%, Philippines 8%, Indonesia 6%, China 6%, Spain 4%, USA 4% and UK 4% [1]. However, till now, no global agency has engaged in collecting the complete statistics on physicians' deaths due to the COVID-19 pandemic among total deaths. In India, the total death rate account of doctors and other medical staff due to COVID-19 has been reported as 0.5% of the total death rate. As per information available with Indian Medical Association (IMA), up to August 2020, 196 deaths were recorded among doctors due to COVID-19 [2],[3]. In other developed European countries, a higher mortality rate has been reported among doctors [4]. Further, as per the latest information published in the Business Standard special report on coronavirus, as of June 16, 2021, the IMA data shows 730 doctors died due to the COVID-19 pandemic in the second wave in India during April-May 2021. The mortality rate in healthcare workers has been witnessed to be on the rise. Many countries have faced a significant surge in mortality among doctors since the beginning of the pandemic. India has continuously shown an upward growth of COVID-19 cases during the first and second waves, however, in comparison with other western countries, India has been seen in a favorable position statistically.

The deaths of physicians represent a drastic and significant loss of the trained healthcare manpower which further affects and damages the healthcare services delivery system. There are many reasons which have contributed to such a significant loss of physicians' lives during the pandemic, such as unpreparedness of the medical staff, inadequate infrastructure, non-availability of protective medicines and PPE kits, lack of life-saving support systems, etc. A safe and secure healthcare services delivery system is required to handle the COVID-19 epidemic-like situations. An effective COVID-19 patient-care system is required which can deliver patient-care services, monitoring and control remotely by avoiding in-person service delivery in hospital wards or without admitting virus-affected patients physically in hospital wards. This type of system can enable the health services providers in early disease detection in virus-infected persons, advise and instruct them to isolate from other people, deliver the required treatment course at their place to follow and monitor their condition remotely from hospital clinics.

An effective remote healthcare service delivery system for monitoring and assessing COVID-19 patients at homes or in hospital wards can utilize and extends the services of medical experts to more patients than in-person patient care. For this purpose, applications of the internet and web-based technologies can fulfill the purpose to provide quality healthcare services in overly populated countries and cities during COVID-19 like pandemics. Internet of Things (IoT) applications have already been started their usage and gaining momentum in healthcare service delivery and management. IoT-based systems incorporate wireless sensing devices which can send and receive data over wired and wireless communication channels. These wireless-sensor devices are very compact and can be worn like a wristwatch, hand-bands, necklaces, rings, etc. These wireless-sensor devices can be integrated with early-warning indicator systems (EWIS) which has the potential to revolutionize and improve the healthcare service delivery system. The Royal College of Physicians London has successfully developed and tested its NEWS2 (National Early Warning Score-2) which has shown significant performance in remote COVID-19 patient information recording. This information is used for risk stratification of COVID-19 patients, however, enough data is not available yet to measure its ability to detect severe COVID-19 cases [5].

The protective and preventive measures have been adopted to control the spread of coronavirus by implementing the social distancing policies, making use of face masks, developing and maintaining basic hygiene habits like frequent hand-washing, sanitizing articles of routine use and checking body temperature regularly, etc. [6-9]. In various countries, complete shut-down was implemented to halt all types of business activities except healthcare and allied services to break the spread-chain of coronavirus. As a result, various industrial, transportation, educational and tourism activities have suffered significantly. The strict measures are essential to reduce the spread of the virus, but they negatively affect the economic growth of the countries, thus, further deteriorating the normal life of people [10]. The applications of wireless sensor network-enabled IoT-based smart wearable devices and technologies can provide improved quality of healthcare services to the COVID-19 infected elderly and physically disabled patients. Such systems are built as multi-component-based systems. The healthcare component can be integrated as a part of the smart home automation module. This module is to be utilized to provide the healthcare service to the patient while staying at home or in other remote location outside the hospitals. It will also reduce anxiety and depression generally faced by COVID-19 patients in isolation at hospital wards. The physicians and medical practitioners can diagnose symptoms, monitor and measure vital health indicators scores remotely from their offices. The automatic smart response systems with user-friendly interfaces enable the patients, especially old age and physically challenged patients, to follow the treatment plan using certain smart home devices connected through the internet such as smartphones, tablets, laptops, etc. Such smart healthcare devices constitute an active framework of computing nodes those act on behalf of the end-user to communicate with the central healthcare service providing server. This type of environment enables and facilitates adequate decision-making on both ends of the system.

This chapter is devoted to a wireless sensor network of smart IoT devices for early detection of COVID-19 patients for remote monitoring, serving, controlling and analysis. This model allows the patient to send his health indicators remotely using internet applications. The smart IoT devices automatically record the patient's health symptoms and other body indicators such as body temperature level, oxygen level (SpO$_2$), pulse rate, breathing, dry cough, etc., and send the information to the designated healthcare server for analysis with automatic machine learning algorithms as well as human expert's analysis. This system model will also build a patient record database to train the machine learning models for a fully automated healthcare service provider model.

The goal of this chapter is to provide basic information on IoT-based technology in the COVID-19 early patient detection, tracking, monitoring and control the spread of coronavirus and review the state-of-the-art IoT-based system architectures, framework platforms and applications of IoT-based health industry solutions combating the COVID-19 in different phases, namely early detection and diagnosis, isolation period and post-COVID recovery period.

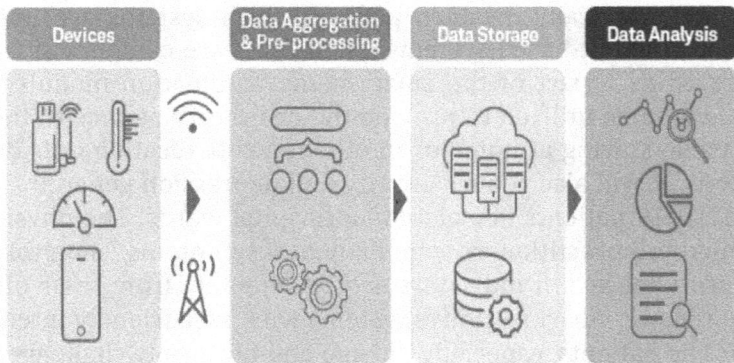

Figure 3.1 Components of the IoT based model healthcare monitoring system architecture for physiological data sensing, collection, processing and storage for analysis purposes.

Figure 3.1 shows the basic IoT-based healthcare monitoring system architecture information workflow. It will be further elaborated with an extension of each component in the next sections.

3.2 RELATED WORKS

The latest published literature in this domain shows the increasing use of technology in remote healthcare monitoring and internet-enabled healthcare services. The healthcare experts are collaborating with technology experts to provide health services through internet services at distance places using IoT service infrastructure. The IoT sensor devices are being used to collect patient's data remotely for diagnosis and corresponding healthcare service delivery. A brief account of the present research trends in IoT-based healthcare service systems is given here. The interested readers of any specific model may refer to the bibliographical references given at the end of this chapter.

In [11], a model for COVID-19 patient detection and monitoring system is presented to collect real-time health symptoms of a virus-infected patient using IoT smart devices and send it to the healthcare server over the internet channels. The patient data parameters are analyzed to predict the coronavirus infectious people using machine learning algorithms such as Support Vector Machine (SVM), Neural Networks, k-mean neighbors (k-NN), and Naïve Bayes, etc. IoT smart device infrastructure is used to detect and monitor both potential and confirmed cases. Simultaneously, the treatment response is also transferred to the concerned patients. This system model also enables the medical experts to understand the nature of the coronavirus from the collected samples by analyzing the symptom trends found in patients. It also creates a database of the relevant recorded data which enables the training of machine learning models for futuristic analysis to make further testing and predictions.

Dry-cough and lung infection create difficulty in breathing which has been found in most corona-infected patients in the second wave of COVID-19. Due to these, thousands of people have died. Till now, no smart device is effectively used to detect cough problems causing breathing difficulty in remote infected patients. IoT-based smart sensor devices can detect cough and respirational symptoms remotely and generate alert warnings to the users to take appropriate protective measures to control the virus spread. The Center for Disease Control and Prevention (CDC) and the World Health Organization (WHO) have issued guidelines, which suggest that people should cover their mouth, nose and face with face masks to avoid touching while coughing and sneezing. In [12], a new smart home monitoring system is proposed that can sense the coughing, sneezing and face touching activities while entering and leaving a room. This model uses radiofrequency technology to sense the motion and action of humans. This model can monitor activities continuously with an overall accuracy of 96% and not only control COVID-19 patients but also support contact tracing by monitoring people's activities within the home. The two main features, coughing and sneezing detection and recording, are remarkable achievements along with other body indicators, whereas the previous system only records audio sensed signals [13],[14].

The face masks have proved very effective to control the spread of virus infection from one person to another. However, this small piece of wearable textile is yet not made smart to collect physiological symptoms. The wearable biosensors can be integrated with face masks for the identification of infected persons. The smart facemasks will be a superior option to use continuously to collect different physiological indicators data from exhaled breath [15].

It is possible to curb the chain of transmission between carriers. Maintaining social distancing has been regarded as a key method worldwide to control the COVID-19 pandemic. However, alerting the people to remind them as well as direct or instruct everywhere and every time is not possible. To make people aware and alert to maintain social distance from each other, IoT wireless sensor technologies can play an important role to support the business, social and day-to-day routine activities outside as well as at home. Digital technologies, which have a proven track record of providing automated quality healthcare service remotely, can also be used to guarantee to maintain social distancing in the surroundings circles. In [16], an end-to-end IoT architecture is proposed which supports the maintenance of social distancing during a pandemic. It is also proposed to decide the short and long-term planning to manage the social-distancing policy using the proposed system model architecture.

In [17], Rao and Vazquez applied machine learning techniques to identify the COVID-19 affected patients. This model learns from the data collected from the users through web surveys using smartphones. In [18] suggested developing standard protocols to identify the coronavirus infected patients in the smart-cities. The AI-based methods can be applied to collect data from the thermal cameras installed in smart cities to identify the coronavirus-infected patients.

In [19], an IoT-based remote health monitoring system is proposed for self-isolated patient collecting their physiological parameter data like SpO_2 and heartbeat rate together with physical location information. This prototype system model uses wireless wearable sensors and a network gateway to acquire and transmit data over internet protocols and employs an application server to store patient data for analysis and visualization.

3.3 WIRELESS SENSORS AND IoT

The availability of cheap-hardware storage devices, high power processors, high-speed wireless data transmission over the internet, compact and cheap wireless smart sensing devices requiring less power, etc., has enabled the use of wireless sensors in many automation applications. One of the major important areas of wireless sensors is the healthcare sector. The use of wireless sensor networks in structural health monitoring is increasing day by day. The fastest-growing wireless technologies are significantly contributing to the progress of structural monitoring systems using wireless sensor networking devices. The use of sensors in health monitoring systems provides a novel technology with compelling advantages in comparison to conventional systems. The wireless technology has reduced the installation and maintenance costs of the hardware infrastructure.

3.3.1 IoT Technology

IoT has emerged as an important data collection and communication transmission technology with existing applications in many areas. It has been known earlier with different names such as pervasive computing, sensor networks and embedded computing devices. The term *IoT system* more suitably describes the use of IoT technology than does the *Internet of Things*. A group of wireless interconnected IoT sensor devices can be used to form an IoT system for any specific purpose which are less frequently accessed devices on the internetwork.

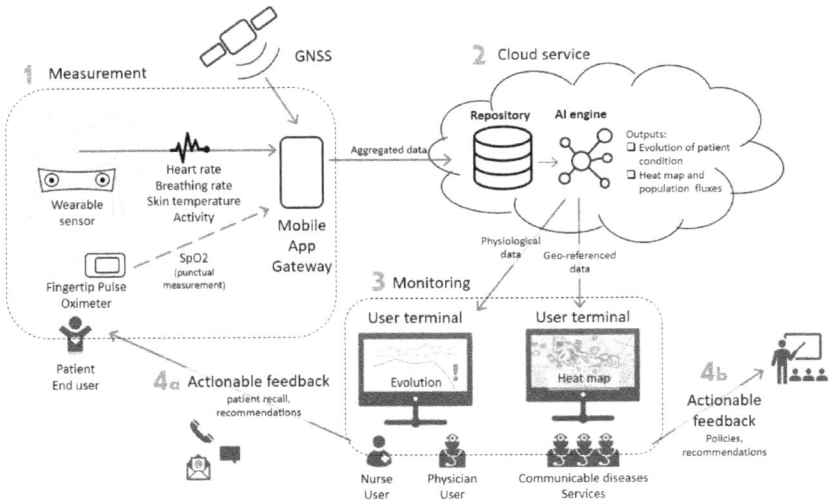

Figure 3.2 Framework for remote patient's automatic physiological data sensing, collection and transmission to data centers for analysis and suitable action. *Source:* https://business.esa.int/projects/como (reproduced with permission).

The availability and use of IoT sensors bring a boom in microelectromechanical sensors, integrated accelerometers, gyroscopes, chemical sensors and other forms of tiny sensors in many applications. The low cost and low power consumption requirements for these sensors enable the developers and researchers to design new application models to deploy in different types of signal processing applications. Figure 3.2 presents the use of the sensing devices to measure different physiological parameters.

The wearable smart sensor network devices are capable to detect and measure physiological indicators and also have the potential to be improved with the use of the emerging new low-cost sensors. These wearable electronics sensors are available in different forms such as contact lenses, wearable textiles, face-masks wristbands, etc., and will improve the data collection by sensing physical and biochemical signals [15].

3.3.2 Working of IoT

An IoT system includes internet-enabled wireless sensor network devices which consist of embedded hardware and software components such as processors, sensors, and communication hardware devices to collect, store, send and receive sensed data on wireless signals in an interconnected environment. IoT sensor devices share the data by establishing a connection with an IoT gateway. These smart devices are designed to work independently without human intervention.

In the healthcare sector, IoT-enabled devices are used in the remote monitoring of patients. The benefits of this are unleashing the potential to provide quality healthcare service to keep patients safe and empowering the medical physicians to perform their job without any worry. It also increases the patient satisfaction as well as interaction with doctors. Furthermore, remote patient monitoring also reduces or avoids hospital ward stay of patients which results in reduced expenditure on treatment costs. IoT is undoubtedly transforming the healthcare sector by smart artificial intelligence-based digital systems which can be trained to look after and take care of remote and far away patients.

3.3.3 Biosensors

Biotechnology and biomedicine category biosensors are used in physiological symptom detection by recognizing biochemical properties of the human respiratory system. These sensors have a significant potential in the identification and diagnosis of multiple disease indicators such as finding multiple comorbidities in elderly COVID-19 patients during the ongoing pandemic [20].

A biosensor generally contains a bioreceptor, a transducer, a processor and a display to detect, process and forward for further transmission of data [21]. The transducer and electronics can be wrapped in complementary metal-oxide-semiconductor materials. Biosensors are used to sense the biomolecular signals from the human respiratory system. Figure 3.3 illustrates the wearable biosensors in personalized healthcare systems.

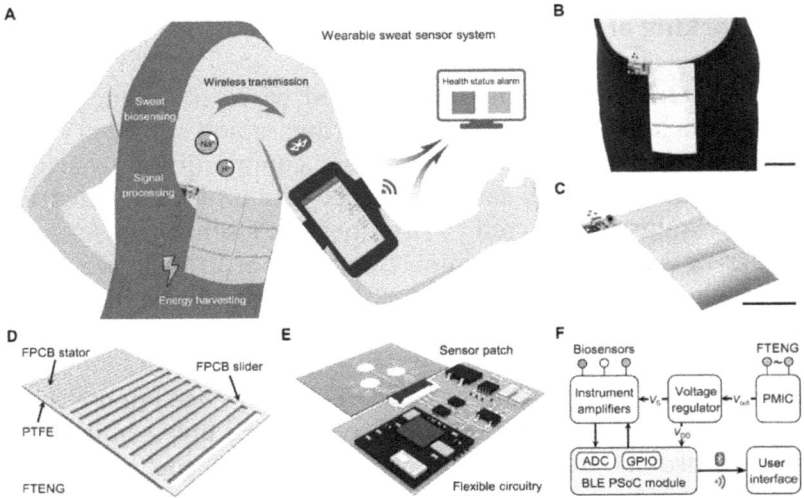

Figure 3.3 In (A) Schematic diagram of wearable biosensor based on human motion energy extraction. It transmits data by Bluetooth transmission to a mobile user interface in real-time health monitoring; (B & C) optical images of FPCB; (D) schematic diagram of FPCB-FTENG with grating slider and an interdigital stator; (E) microfluidic-based sweat sensor patch interface; and (F) represents the system level block diagram for power management, signal transduction, processing, and wireless transmission of data through biosensors to the user interfaces. *Source: California Institute of Technology* (reproduced with permission).

The main task of biosensors is producing data results and to allow fast testing at the exact point of interest. An energy-efficient and highly robust battery-free wearable sensor-based platform has been proposed in [22] which can recharge itself by extracting energy from the body motion. It uses its flexible printed circuit board (FPCB)-based free-standing triboelectric nanogenerator (FTENG) which displays high power output of ~416 mWm^{-2}. This biosensor platform also multiplexes sweat biosensors which can transmit data wirelessly to the user interfaces using Bluetooth when the sensor is on the body [22].

3.3.4 IoT in Remote Health Monitoring

To design an IoT-based remote healthcare monitoring system, different types of biosensors are required to be integrated with the internet infrastructure components to transmit collected data. The effectiveness of the IoT-based healthcare monitoring system can be ensured by incorporating patient identification module, data collection and transmission of recorded physiological parameters to the data analysis server for processing and analysis to predict the results. It also generates an alert message for medical caregivers, hospital authorities and civil administration to initiate appropriate action and inform the people to follow the standard health protocols. The IoT-based remote healthcare model will comprise different layers containing different wireless sensors, actuators, microcontrollers, processors and transducers, which are used to transmit the patient data through the wired or wireless telecommunication channels to the cloud or local server for processing and analysis.

3.4 REMOTE HEALTHCARE MONITORING IN COVID-19 PANDEMIC

The world has witnessed the death of thousands of COVID-19 infected people since its outbreak in January 2020. People can easily get affected with COVID-19 from other infected people. It is almost impossible to detect a COVID-19 affected person until he/she is not tested. The infected people with no visible virus symptoms are more prone to spread the virus, as it spreads by direct contact from person to person. The virus spreads rapidly through human respiratory droplets when infected persons cough or sneeze. It also spreads from the infected surfaces when we touch these surfaces and subsequently touch our eyes, mouth or nose without sanitizing or disinfecting the hands [23]. This pandemic poses a dangerous health risk, especially for elderly people with other health comorbidities [24]. Overall, this pandemic has led to a major crisis. Many countries have struggled to manage their near collapsed healthcare systems. During the peaks, most hospitals were full of their capacity with COVID-19 patients and thousands of people struggled to find a hospital bed. A major problem erupted when the virus started to spread among the doctors and other healthcare professionals in these hospitals. It created a major health risk for patients admitted to hospitals and hospital workers. As per World Health Organization (WHO), the population of people aged 60 and above is continuously growing. Providing quality healthcare to all elderly people while admitting them physically in under-resourced hospitals in a pandemic presents a serious threat to the life of all people. Therefore, the governments, healthcare workers and technology people should device and adopt alternate healthcare techniques. The IoT technology is a superior option to provide quality health services by home hospitalizing the COVID-19 patients. Home hospitalization will provide benefits in two ways: first, the patient will stay in own home environment which will positively affect the use of medicines, thus, avoiding the movement to and from hospitals, receiving continuous monitoring and living in the home comfort. Secondly, the treatment cost will be lowered by avoiding hospital bed charges, room rent, transportation charges, etc.

The evolution in IoT technology is providing great mechanisms and opportunities to build smart objects and intelligent devices that may have a tremendous ability to collect, share and analyze data between other devices [24]. The IoT technology is widely accepted all over the world and is being used in many significant applications including healthcare services for monitoring patient health remotely.

3.5 REMOTE COVID-19 DETECTION

Wearable wireless sensors automatically detect, collect and send COVID-19 patients' physiological indicators such as respiration rate, heart-beat rate, pulse rate, motion activity, blood oxygen saturation level, dry-cough analysis and stress level. Table 3.1 shows the sensor metrics for virus infection indicators sensed in the person wearing the sensors. These physiological data variables are forwarded to the healthcare server by the sensing device for further processing and analysis to predict whether a person is virus-infected or not [25], [26] and [27].

Table 3.1 Health indicator metrics used in IoT sensor-based measurement

Sensor Device	Heart Beat	Respiratory Rate	SpO$_2$	Motion Activity	Skin	Location	Cough	Resting Heart Rate
Accelerometer		√		√		√		
Temperature					√			
GPS						√		
ECG	√	√						√
Oxygen			√					

PPG	√	√	√	√				√
Micro-phone							√	

3.6 EARLY COVID-19 DETECTION

As vaccines are coming slowly and underway in the market, it becomes more important to find ways to detect the infected people as early as possible to avoid the spread of the virus among others. It is seen that the wearable wireless IoT-based sensor technology can detect, monitor and predict virus infection incidents using sensor devices to measure health condition indicators such as heart-beat rate, body temperature, blood oxygen saturation, coughing, breathing patterns and other human respiratory symptoms [28].

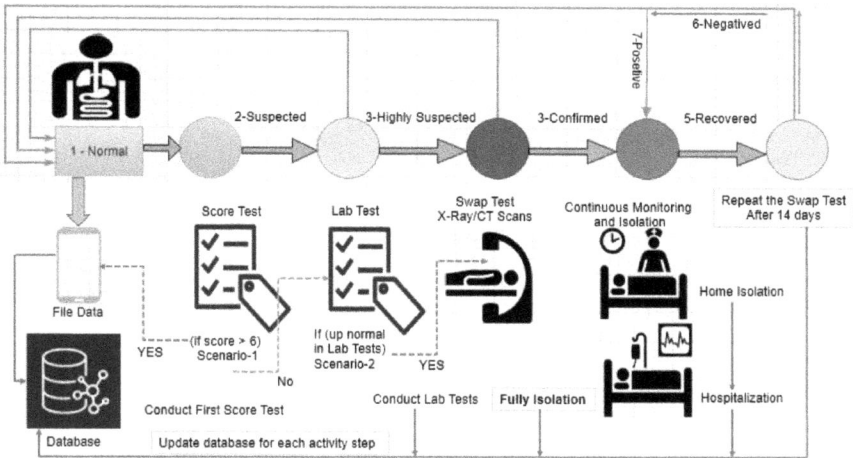

Figure 3.4 Patient data collection network flow diagram of remote COVID-19 patient symptom data processing and verification system model at the base station. Starting from the very early stage to COVID-19 detection and afterwards health caring and monitoring is depicted in numbered steps.

Historical trends in available COVID-19 data show that the rapid disease control is possible by early detection of infected patients and isolating them from others. The only effective and viable way to avoid and curb the spread is early detection of the COVID-19 symptoms in affected persons when no or limited effective vaccines are available. A physical person-to-person sample collection can further result in the transmission of the virus among the masses. The very first step that can be taken in this direction is to detect and verify the virus infection remotely using IoT sensors like automated means.

Never before was a virus detection and verification procedure so critical in the medical history. In Figure 3.5, it has been shown that the more extensive the early detection and testing of COVID-19 is conducted, the lowest is the overall mortality rate in that country. It confirms the importance of early testing and detection to curb the spread of the deadly COVID-19 virus. It is also inevitable from the figure that Korea and Germany have conducted a greater number of COVID-19 tests since the outbreak of the pandemic and have one of the lowest death rates corresponding to other countries.

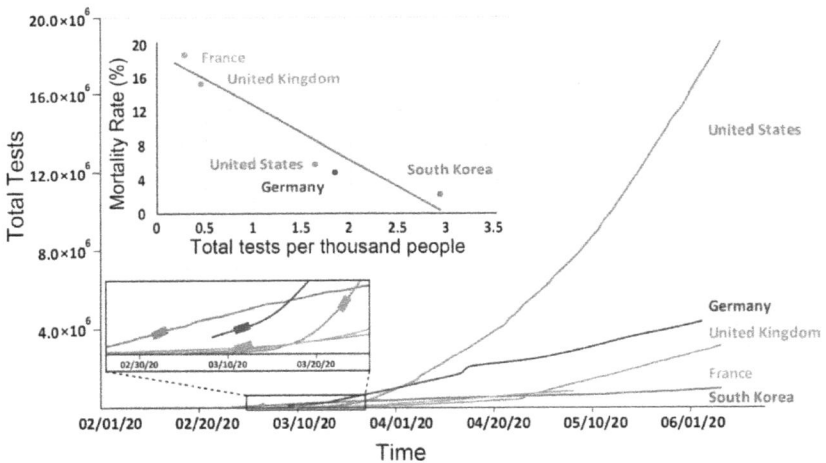

Figure 3.5 The daily testing of COVID-19 in five countries with similar resources is shown. The inset figure shows the death rate among all COVID-19 confirmed cases up to June 7, 2020, vs. some early detections per thousand population conducted during March 4-26, 2020. The inset linearly fit data indicates a negative correlation between the early detection and death rate [29]. *Source: WHO* (reproduced with permission).

Therefore, it becomes important to safeguard the medical caregivers first and simultaneously adopt safe and effective protective measures to safeguard the general public in the pandemic. The use of IoT technology can revolutionize the healthcare sector following standardized secure procedures to combat the COVID-19 pandemic [30].

3.7 IoT REMOTE HEALTHCARE SYSTEM ARCHITECTURE

The IoT-based remote monitoring system will have a minimum of three layers.

- Data Acquisition Layer
- Data Distribution Layer
- User Application Layer

The system will detect the potential virus-infected person by reading physical indicators and geographical location. The Data Acquisition Layer is responsible to collect the patient's physiological parameters using IoT sensor technology [31]. The Data Distribution Layer will allow the patient's physiological parameters distribution among the different system modules to process and analyze the data at the cloud server or on the local server. This layer will also alert and warn the concerned caregivers in the case of any emergency conditions.

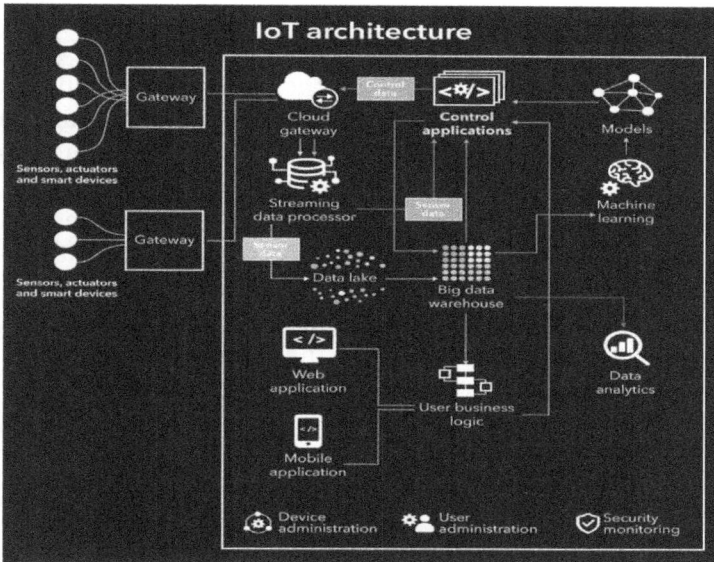

Figure 3.6 Conceptual detailed architecture of IoT-based remote healthcare monitoring system.

The third Application Layer will be used to allow the users to access the system functionality on user (patient and healthcare staff) smartphone devices, desktops or laptops computers, smart-tablet notebooks to analyze the processed results, system messages and other system administration and configuration management tasks [32]. Figure 3.6 shows the conceptual system architecture of the proposed system architecture to meet the expected requirements.

Data Acquisition Layer will acquire the IoT sensor data responses to be collected from the patient's health symptoms. The patient will have to wear or attach the IoT sensor kit to his body to read his/her vital body indicators. The acquired data will be transmitted over the internet lines using transmission control protocols through the gateway to the data processing server. The Data Distribution Layer will act like a message passing between the individual data generating sensor devices and data consuming processes in the other two layers.

The Application Layer has three sub-layers modules:

- User Interface module
- Data Processing module
- External Communication module

The User Interface module allows the access of all applications through the user interface on desktops, laptops and smart mobile devices [33]. All database activities such as data processing, analysis and storage management are handled by the Data Processing module. The External Application Layer supports the other institutional information management processes such as patient health records management and enterprise resource planning (ERP). These three sub-modules can exchange data by message passing to each other. Embedded software agents manage the inter-device interaction by customizing remote feature configuration in each layer.

3.8 DATA ACQUISITION LAYER

The Data Acquisition Layer has two basic modules – the Device module and the Gateway module. The Device module works as the sensor data collector, whereas the Gateway module is responsible for the integration of the sensor devices to higher layer modules [33].

3.8.1 Gateways

The Gateways work to establish connections with wearable sensor kits at remote COVID-19 patient's places executing the processes on the Application Layer through the Data Transmission Layer. Two types of radio-frequency protocols are used to send and receive messaging – low-power to connect the wearable sensor devices and message passing network through Wi-Fi or a mobile network [34].

3.8.2 Gateways Configuration

Two types of Gateways are used: the fixed base-station Gateways are used in the hospital's premises, and the second client application is used on the patient's smartphone device if the patients are home isolated. The Gateways are registered and activated to start establishing connections with wearable sensor devices at remote sites [34]. The patient's privacy and data are protected, and data security is implemented by Gateway protection with advanced encryption standard protocols. Transport Layer Security is enforced during message-passing between the Gateways and data center.

Each Gateway is configured assigning a unique identifier to establish connection only with limited wireless sensor nodes. In this way, the Gateway in the smartphone devices can only make connections from the wearable sensor devices attached with respective COVID-19 patients, whereas the Gateways at hospital wards can make multiple connections with multiple smartphone devices. Gateways are also intelligent devices that can control the energy consumption of wearable sensor devices by optimizing the energy consumption. For example, when the radio signal is always on, the Gateway only permits the wearable device to turn on when it is required to transmit data. This method saves energy consumption in sensor devices without any data transmission. The fixed Gateway at hospital wards can also be deactivated due to any reason. A deactivated Gateway will not communicate with wearable sensors and will not allow reaching the cloud services in higher layers.

The wearable sensor devices transmit patient's health data to Gateways. The data can be visualized on the receiver-side GUI applications installed on smartphones, tablets and laptops. Further, these devices transmit this data to the message-passing agents running on the cloud server or local applications servers. Due to both-way communication, the Gateway can also update the sensor devices' configurations and manages the flow of configuration messages using low-energy transmission protocols. The Gateway can be configured to invoke the sensor device at fixed intervals to optimize the battery power to extend the battery life cycle.

3.8.3 Physiological Sensors

To design a wearable sensor kit to collect physiological symptom data of the COVID-19 patients, low-cost suitable tiny sensors are used. Different type of sensors are required to collect different types of physiological symptoms.

Table 3.2 Wireless sensors for different physiological parameters

Sensor	Purpose	Technology	Composition	Calibration
MAX30100	Heart Rate/Pulse	Infrared, Red light frequency to determine the %age of HB in blood	2 LED, photo-detector, en-hanced optics, and low-noise analog signal processing	Programma-ble from 200 μs to 1.6 ms to optimize measurement accuracy
MAX30205	Body Tem-perature	Uses High-Resolution ADC to con-vert temper-ature in digi-tal form	USB-I^{2}C with display	As per ASTM E1112(0.1 °C) meet clinical thermo-metry speci-fication
SW-420	Cough and Motion	Doppler Ra-dar, CW-Radar & vi-bration de-tection	LM393	With onboard potentiome-ter for sensi-tivity thresh-old selection
MAX30100	Blood Oxy-gen (SpO$_2$)	Infrared, Red light frequency to determine the %age of	2 LED, photo-detector, en-hanced optics, and low-noise analog signal	Programma-ble from 200 μs to 1.6 ms to optimize measurement

		HB in blood	processing	accuracy
MAX86150	ECG	LED photo-detector	Low noise electronics	USB-I²C on 1.8 V battery
ESP8266	Wi-Fi Connectivity	Integrated TR switch, PLL regulator, 32-bit CPU	Full TCP/IP stack and microcontroller connectivity	Wakeup and transmit data packets in < 2 ms

For example, to detect and measure fever, a body temperature measurement sensor is required. For blood oxygen level and respiratory rate measurement, a SpO_2 sensor is required which can also measure pneumonia, anemia, asthma, etc. Similarly, for cough and breathing variability detection, a motion sensor can be selected. Table 3.2 gives the basic sensor hardware information required to build a COVID-19 remote patient monitoring system. The choice of sensor types depends on our requirement to detect the vital physiological parameters as given in the table, i.e., body temperature, blood oxygen saturation level, dry cough and motion, heart rate and Wi-Fi connectivity.

The wearable sensor kit requires a low-power operated microcontroller radio transmission transceiver. So, it can be operated on Bluetooth short wave radio frequency protocols to connect sensors with Gateways. Also, the IEEE 802.15.4 protocols can be used which allows the use of customized protocols to reduce power consumption and to enhance battery life. After connecting the nearest Gateway using appropriate radio frequency protocols, the sensor kit can send the collected data parameters to the remote system for analysis.

Each wearable sensor kit has a unique identifier and device status codes, i.e., a) in use, b) free & available, c) clearing and d) out-of-service. It will only be active when worn or attached to a patient. It is disinfected when returning to the hospital corresponding to the clearing status. When the kit is ready, it becomes available. In any technical fault condition, its status turns to 'out-of-service', thus, requiring maintenance. It can be controlled with administrative privileges making a direct connection by any specific GUI application installed on a smartphone or tablet or laptop computer.

3.8.4 Physiological Data Collection

The wearable wireless sensor remote monitoring kit collects patient's data in analog and digital form. This data is filtered and transformed into a set of consolidated data values calculating its mean or median within a specific time interval. These values are used to compute the individual score in different classes for each sensor node. A total score is computed to assess the current clinical status before sending these data values [35].

The wireless sensors module sends three types of messages:

- Periodic Messages – discrete signal values of each sensor.
- Alert Messages – when the composed data values indicate deterioration in the patient's health condition.
- Conditional Messages – if...then condition-based message when sharp ups or downs are detected in continuous data trends.

In addition, when the battery level goes down at a predetermined level of 5% or 10% remaining, a battery recharge or replacement message is also flashed in the controlling dashboard as well as sent to the patient's mobile phone interface. Precautions are taken to use the minimum energy to extend the battery life using small text messages and only when it is required.

3.8.5 Sensor-kit Configuration

Embedded software is used for configuring with various features to handle signals from sensor nodes, consolidating signal values and extracting clinical status parameters from the acquired sensor signals [35]. These signal parameters are used to trigger alert messages, setting device connections and communicating the previously given messages. The software is automatically installed on the sensor module when it is turned on. Each sensor module has a unique identifier without the patient's identity [36]. Further, the wearable sensor kit is decomposed into two types of parameter assessments: a) operational parameters assessment and b) health indicator parameters assessment.

a)   Operational parameter assessment is concerned with regulating the signal reading variability of each sensor, radio and processor working modes, reading of sensor's activation and deactivation states, transmission protocol control, connection time-outs, message size buffering, checking and alerting battery levels, message encryption, other hardware tuning parameters, etc. The default mode setting for message passing is radio communication which consumes less energy. When the sensor-kit is connected, it is auto-configured using its identifier key. Initially, a parameter is composed in the "Key-Value" pair property which is different from default values. The default value templates are selected to use according to the initial parameter values.

b)   The health parameter assessment is conducted to extract early warning indicators which are extracted from each sensor signal value. Each sensor in the wearable sensor kit collects different physiological health parameters from the body of attached patients. These sensor parameters are categorized into eight different classes. Each class has a range of continuous signal values in the minimum to maximum threshold range at each level of signal value. Each class has also an associated parameter score. Each sensor score value is matched with the corresponding class and given a score accordingly. Each sensor score received at a specific time interval is aggregated to compute the composite score. This score is classified into different categories which represents the clinical status of the COVID-19 patients such as mild, moderate, severe and critical. The time interval for each category of the score is measure in seconds, and it is also used to regulate the message transmission interval from wearable sensor kit to gateway. Depending upon the health condition status code of the patient, the message time interval can be decreased or increased. The wearable sensor kit duty cycle is required to adjust to optimize the battery power consumption, i.e., message sending time interval can be increased 5 to 10 minutes for non-critical or mild condition patients which allows a longer monitoring duty cycle [35], [36].

68   *Emerging Trends in Wireless Communication*

Table 3.3 Individual IoT sensor measurement of different physiological parameters of a patient

Clinical Health Status Checking									Sensor Data Calculation	
Class	1	2	3	4	5	6	7	8	Statistical Calculation	Sliding Window size
Respiratory Rate	8	-	11	20	-	24	99	-	Median	10	5
SpO$_2$	91	93	95	99	-	-	-	-	Median	5	0
Heart Rate	40	-	50	90	110	130	99	-	Median	5	0
Temperature	35.0	-	36.0	38.0	39.0	99	-	-	Mean	20	10
Cough											
Systolic BP	-	-	-	-	-	-	-	-	Mean	20	10
Result Score	3	2	1	0	1	2	3	-			

The sensor kit software module saves the previous clinical status to compare it with the new score, and if is higher than the previous, an alert message is generated. The clinical status condition is required to arrange in sequential order so that appropriate messages can be triggered automatically. The physiological score can also be arranged according to a specific age group. Therefore, for older adults, separate parameter assessment criteria and different message-passing time intervals should be selected. Any subtle changes in health status parameters for older age group patients require more attentive protocols than younger adults.

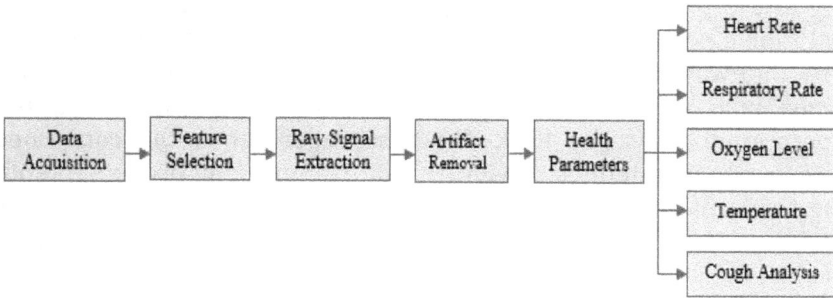

Figure 3.7 A data-flow diagram for data acquisition to decision making based on the data-centric automatic prediction of infection diagnosis in COVID-19 affected persons.

Any drastic changes in the older group during any specific time interval would be considered abnormal, and an alert message will automatically change the clinical status to attract more attention from the medical staff.

If required, the sensor-kit configuration can monitor and reconfigure during its operation. When it sends a message to the Gateway, its waits for an acknowledgement message [37], [38]. If the acknowledgement code indicates the messages are queued in to deliver from Gateway to sensor-kit, it will receive the reconfiguration messages and the changes will apply after processing the next cycle of the clinical status parameter. In this way, the acknowledgement status flag can be set to turn off the radio transmission to save the battery life.

3.9 DATA DISTRIBUTION LAYER

The Data Distribution Layer distributes the generated data from the sensor-kit devices by message passing between different components of the system. An external auxiliary API module also collects sensors data from other devices. This layer has two components: message passing and external API [39], [40].

3.9.1 Message Passing Component

The message passing module receives the data generated by sensor-kit devices, which is distributed among the data consumers according to the pre-subscribed policy. Each specific generator generates data for a specific data receiver identified with its unique identity. The Gateway receives the sensor identity and generates message types such as periodic messages, alert messages, conditional and operational messages [39]. The messenger sends the concerned message to its intended consumer. For example, any process intending to receive a periodically monitored message will receive them based on the unique identifier of the sensor and its location. Then, healthcare staff receives the generated data they are supposed to receive for analysis of health status update. Another process collects and stores the received data into a database for further analysis. This process continues to trigger messages in parallel to change and update clinical status. At cloud or local server-side, the received data is integrated into the hospital's medical record management system. The processed information is displayed on the ward's controlling dashboard, medical attendants' smartphones or internet web portals for visualization [40]. The patients also receive the same information on their smart mobile phones or computers. Continuous monitoring provides regular updates generated from the patient data collected by IoT devices.

3.9.2 External API

The external web-API software module allows accessing third-party sensor devices such as smartphones and smartwatches. This module collects data from other sensors and converts it in the compatible form to use by this system. However, its use will increase the latency and dependency of network connections as well as increase the network traffic flow [40].

3.10 APPLICATION LAYER

The Application Layer has the following three main parts:

- User application
- Data processing
- External application

This layer integrates different modules as well as aggregates, stores, presents and processes the acquired data. Each part of this layer comprises several processes as given in the following sub-sections:

3.10.1 User Applications

This application interface provides system access to all end-users for visualizing and interacting with the health service activity pipeline. Secondly, system administration and database management tools are also accessed here. This application interface provides access to COVID-19 patient outside the hospital through the Gateway [41]. In addition, the patient himself can also send his health symptom to the system online such as fever, temperature, cough, shortness of breath, headaches fatigue, sore throat, etc., in addition to the data automatically collected by the sensor devices. The display dashboard is also accessible through this application for patient data visualization in hospital wards on different types of wired or wireless devices [42]. Each patient's physiological health indicators are displayed indicating clinical health status. The administrative tools are used to manage the data model and system parameter configuration. Only system monitoring parameters are allowed here to access the system management [43].

3.10.2 External Application

Some external applications are required to access the system and also provide access to other external applications to share the data and other resources in sister institutions and government agencies. Some other corporate applications that belong to health care also require to access the monitoring system to share useful information such as laboratory reports and medical image scan reports. The patient's data is required to be shared among the authorized health agencies and research organizations for future investigations. Therefore, external applications are required to allow interaction directly or indirectly with the healthcare monitoring system through web service interfaces. For this purpose, the institutional consent management system manages the authorization for data collection, storing and sharing patient health records [44].

3.10.3 Data Processing

The collected data from each wireless sensor-kit is processing by the data processing module in the Application Layer. This module has four main components as event management, stream processing, machine learning and storage management. These components are used to perform various data processing and analysis techniques to extract the predictable information [45].

The event management part control and manage the event requests coming from the User Application Layer, external applications and from message-passing processes, as shown in Figure 3.8.

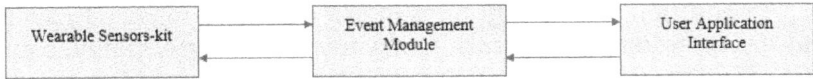

Figure 3.8 IoT Device-Event-Activity management diagram between data-sensors and user applications.

The periodicity of the received messages is checked based on event frequency configuration. The working status of each wireless sensor kit is continuously monitored by the supervisor of a particular patient. If the system is not receiving any message from the sensor module, the system triggers an alert warning message. If any change has occurred, the reconfigure message is forwarded to that module. The stream processing handles the sensor data streaming and continuously updates the patient health status. This is a very important event that helps in managing required resources in any prevailing situation.

The data processing component is responsible for data transformation and feature extraction from the in-flowing data. Machine learning algorithms are used to classify the data clustering to assist in outcome prediction and decision-making. The stored data will be helpful to train the machine learning models to extract the temporal data patterns in the collected patient's data. This data will provide useful knowledge base by data mining for future decision-making to minimize the disease effects and mortality rate in the general population [18], [30], [45].

3.11 POTENTIAL CHALLENGES

There are several issues and challenges in internet-enabled IoT-based healthcare remote monitoring systems. This digital healthcare system uses smart technologies like IoT wireless sensors and big data which are expected to seamlessly allow the connection of patients and different health-service provider systems [39]. These smart healthcare systems are increasingly connected to the internet via different types of wearable compact sensor technologies for around-the-clock real-time healthcare monitoring. The percentage of the population in developed and developing countries adopting wearable remote healthcare technology devices is rapidly growing. However, several challenges need to be addressed while adopting these internet-based caregiver systems. Here, some major challenges, which can hinder the wider acceptance of wireless sensors based remote IoT healthcare systems, have been discussed.

Privacy and security – IoT sensor devices send and share user medical records over the internet. If proper protective measures are not applied, it can pose a threat to user's privacy and data security over the internet. The wireless devices collect patient data, aggregate it, process it and transfer it to the cloud or local base server. These devices on the physical layer are vulnerable to tag cloning, spoofing, RF jamming and cloud polling, by which network data traffic can be diverted to other commands. The direct connection attack is fired to locate the target IoT device to disrupt the service delivery [46], [47].

Denial of service attack – This type of attack is used to affect the healthcare system and patient safety. These attacks can be made ineffective providing duplication of resources, but in the healthcare domain, this method is not always effective as many devices are attached with the implanted life-saving machines. The immediate detection of potential security risks always remains a big challenge due to a large number of vulnerabilities in the hardware and software modules. The non-secured internet transmission protocols increase the risk factor. The wearable sensor kits are vulnerable to proliferation such as embedded sensors and implanted medical devices. Due to a lack of security standards, wireless wearable devices are more vulnerable to all kinds of attacks.

Some patients may not be interested to make public their medical records due to the sensitive nature of health data like in HIV or cancer cases. Therefore, the medical healthcare system should comprise health-data confidentiality. The privacy issues stem from the fear of hacking of e-mails, bank accounts, theft of personal documentation and personal details. These risks increase when patient data is shared among several applications [48], [49]. In such cases, it is replicated at several places and chances of its misuse become high.

Wearable sensor kits operated with battery power - The battery weight and size are proportionally large to power capacity and life. Battery weight and physical structure should be small to provide comfort and easy handling. Long-life compact batteries come with high costs. Therefore, the system should be energy-efficient and work on optimized power consumption. In some regions, the internet and network bandwidth availability are major limitations. Some people may not have a smartphone or tablet or laptop computer and may also not be in a position to afford the internet expenditure.

Some other sensitive issues may affect the effective use of IoT-based remote healthcare monitoring systems during a pandemic such as accurate data recording collection, analysis and prediction. Complete data protection and security are not possible in a digitally interconnected world until the security signatures are not frequently updated with utmost care. The cyber-attacks are more difficult to be combated in highly connected and dynamically distributed networks. However, to fight with COVID-19 pandemic, which is a threat to humanity all over the world, the benefits are bigger than the limitations and risks in the IoT-based healthcare systems [50].

3.12 FUTURE SCOPE

There is a bigger scope of IoT devices in remote health monitoring. Soon, the IoT will dominate the healthcare sector which will utilize mostly smart devices and embedded applications. This emerging technology domain consists of several interconnected *"things"*, items that can connect anything, anybody, anytime, anywhere and everywhere on any network. According to Baker's Hospital review report [51], in the coming years, nearly 3 out of 5 hospitals will use IoT devices in their medical services, 73% will monitor patients using IoT devices, 87% of hospitals are planning on IoT implementation, and up to 57% costs are reduced in their service facilities. However, patient privacy protection and data security are big challenges in IoT-based healthcare services delivery. The IoT healthcare sector was valued at US $28.42 billion in 2015, which is now expected to grow to more than $ 300 billion by 2022 and is projected to touch the US $337.41 billion by 2025. Figure 3.9 shows the trends in IoT technology in the healthcare sector.

Global IoT in Healthcare Market (2014-2025)

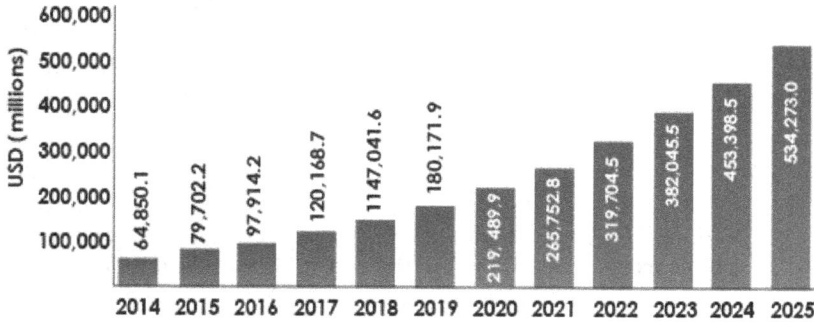

Figure 3.9 Estimated future growth scope of IoT technology in healthcare. *Source: [52]* (reproduced with permission).

The IoT technology will provide the affordable and around the clock quality healthcare at home which will eliminate the various reasons that lead to stress and tension to both the patients and relatives. The wearable IoT devices will automatically monitor the patient's condition and continuously inform the caregiver doctors. There are tremendous growth opportunities for different types of skill sets and different types of technologies will also emerge to serve the society effectively.

3.13 CONCLUSION

The worldwide spread of COVID-19 has taken more than 41 million human lives. The use of IoT technology in remote health monitoring is safe, effective and cheap, and it can assist the medical experts to provide dedicated healthcare service. The IoT-based wearable healthcare monitoring devices enable doctors, patients and family members to monitor the patient with their mobile phone applications while staying at home. It will be used to reduce the rush of affected people in the hospital during pandemic-like situations. The prototype models of the system have shown reliable performance and intelligence to provide safe and low-cost quality healthcare. The three-layer architecture of the wearable IoT-based body sensors, web-API and responsive user application interfaces will be helpful to alleviate the stress on people, doctors, medical authorities and patients. The layered architecture of the wearable sensors will assist in the collection of the vital physical body signs data of the patients like temperature, oxygen saturation levels and cough analysis, along with the geographical location data of the patients, to the medical authorities. The application device interfaces will assist in the data analysis, monitoring and system configuration management, among other functions.

References

1. E. B. Ing, Q. Xu, A. Salimi, and N. Torun, (2020), "Physician deaths from corona virus (COVID-19) disease," *Occup. Med. (Chic. Ill).*, vol. 70, no. 5, pp. 370–374.
2. "IMA says nearly 200 doctors in India have succumbed to COVID-19 so far: requests PM's attention." [Online]. Available: https://economictimes.indiatimes.com/news/politics-and-nation/ima-says-nearly-200-doctors-in-india-have-succumbed-to-covid-19-so-far-requests-pms-ttention/articleshow/77430706.cms.
3. "Coronavirus in India: how the COVID-19 could impact the fast-growing economy," [Online]. Available: https://www.pharmaceutical-technology.com/features/ corona-virus-affected-countries-india-measures-impact-pharma-economy/.
4. P. Fusaroli, S. Balena, and A. Lisotti, (2020), "On the death of 100+ Italian doctors from COVID-19," *Infection*, vol. 48, no. 5, pp. 803–804.
5. E. Carr *et al.*, (2021), "Evaluation and improvement of the National Early Warning Score (NEWS2) for COVID-19: a multi-hospital study," *BMC Med.*, vol. 19, no. 1, pp. 1–16.
6. J. Liu *et al.*, (2020), "Community transmission of severe acute respiratory syndrome coronavirus 2, Shenzhen, China, 2020," *Emerg. Infect. Dis.*, vol. 26, no. 6, p. 1320.
7. J. Howard *et al.*, (2021), "An evidence review of face masks against COVID-19," *Proc. Natl. Acad. Sci.*, vol. 118, no. 4.
8. M. U. G. Kraemer *et al.*, (2020), "The effect of human mobility and control measures on the COVID-19 epidemic in China," *Science*, vol. 368, no. 6490, pp. 493–497.
9. X. Wang *et al.*, (2020), "Impact of social distancing measures on coronavirus disease healthcare demand, central Texas, USA," *Emerg. Infect. Dis.*, vol. 26, no. 10, p. 2361.
10. A. Haleem, M. Javaid, and R. Vaishya, (2020), "Effects of COVID-19 pandemic in daily life," *Curr. Med. Res. Pract.*, vol. 10, no. 2, p. 78.
11. M. Otoom, N. Otoum, M. A. Alzubaidi, Y. Etoom, and R. Banihani (2020), "An IoT-based framework for early identification and mon-

itoring of COVID-19 cases," *Biomed. Signal Process. Control*, vol. 62, p. 102149.

12. E. Miller, N. Banerjee, and T. Zhu, (2021), "Smart homes that detect sneeze, cough, and face touching," *Smart Heal.*, vol. 19, p. 100170.

13. A. Jasmine, and A. K. Jayanthy, (2020), "Sensor-based system for automatic cough detection and classification," *Test Eng. Manag.*, vol. 83, pp. 13826–13834.

14. T. Drugman *et al.*, (2013), "Objective study of sensor relevance for automatic cough detection," *IEEE J. Biomed. Heal. informatics*, vol. 17, no. 3, pp. 699–707.

15. H. C. Ates, A. K. Yetisen, F. Güder, and C. Dincer, (2021), "Wearable devices for the detection of COVID-19," *Nat. Electron.*, vol. 4, no. 1, pp. 13–14.

16. P. Coupé, J. V. Manjón, E. Gedamu, D. Arnold, M. Robles, and D. L. Collins, (2010), "Robust Rician noise estimation for MR images," *Med. Image Anal.*, vol. 14, no. 4, pp. 483–493, doi: 10.1016/j.media.2010.03.001.

17. A. S. R. Srinivasa Rao, and J. A. Vazquez, (2020), "Identification of COVID-19 can be quicker through artificial intelligence framework using a mobile phone-based survey when cities and towns are under quarantine.," *Infect Control Hosp Epidemiol*, pp. 1–5.

18. Z. Allam, G. Dey, and D. S. Jones, (2020), "Artificial intelligence (AI) provided early detection of the coronavirus (COVID-19) in China and will influence future Urban health policy internationally," *AI*, vol. 1, no. 2, pp. 156–165.

19. R. Priambodo, and T. M. Kadarina, (2020), "Monitoring Self-isolation Patient of COVID-19 with Internet of Things," in *2020 IEEE International Conference on Communication, Networks and Satellite (Comnetsat)*, pp. 87–91, doi: 10.1109/Comnetsat50391.2020.9328953.

20. F. Cui, and H. S. Zhou, (2020), "Diagnostic methods and potential portable biosensors for coronavirus disease 2019," *Biosens. Bioelectron.*, vol. 165, p. 112349.

21. A. Hierlemann, and H. Baltes, (2003), "CMOS-based chemical microsensors," *Analyst*, vol. 128, no. 1, pp. 15–28.

22. Y. Song *et al.*, (2020), "Wireless battery-free wearable sweat sensor powered by human motion," *Sci. Adv.*, vol. 6, no. 40, p. eaay9842.

23. W. Guan *et al.*, (2020), "Clinical characteristics of coronavirus disease 2019 in China," *N. Engl. J. Med.*, vol. 382, no. 18, pp. 1708–1720.

24. L. Wang *et al.*, (2020), "Coronavirus disease 2019 in elderly patients: characteristics and prognostic factors based on 4-week follow-up," *J. Infect.*, vol. 80, no. 6, pp. 639–645.

25. B. Johansson, S. Jain, J. Montoya-Torres, J. Hugan, and E. Yücesan, "Effective Real-time Allocation of Pandemic Interventions."

26. O. S. Alwan, and K. P. Rao, (2027), "Dedicated real-time monitoring system for health care using ZigBee," *Healthc. Technol. Lett.*, vol. 4, no. 4, pp. 142–144.

27. R. C. C. Dantas, P. A. De Campos, I. Rossi, and R. M. Ribas, (2020), "Implications of social distancing in Brazil in the COVID-19 pandemic," *Infect. Control Hosp. Epidemiol.*, pp. 1–2.

28. N. Al Bassam, S. A. Hussain, A. Al Qaraghuli, J. Khan, E. P. Sumesh, and V. Lavanya, (2021), "IoT based wearable device to monitor the signs of quarantine remote patients of COVID-19," *Informatics Med. Unlocked*, vol. 24, p. 100588.

29. P. Pokhrel, C. Hu, and H. Mao, (2020), "Detecting the coronavirus (COVID-19)," *ACS sensors*, vol. 5, no. 8, pp. 2283–2296.

30. H. Bolhasani, M. Mohseni, and A. M. Rahmani, (2021), "Deep learning applications for IoT in health care: A systematic review," *Informatics Med. Unlocked*, p. 100550.

31. A. Hajizadeh, and E. Monaghesh, (2021), "Telehealth services support community during the COVID-19 Outbreak in Iran: Activities of Ministry of Health and Medical Education," *Informatics Med. Unlocked*, vol. 24, p. 100567.

32. A. I. Paganelli *et al.*, (2021), "A conceptual IoT-based early-warning architecture for remote monitoring of COVID-19 patients in wards and at home," *Internet of Things*, p. 100399.

33. B. Pradhan, S. Bhattacharyya, and K. Pal, (2021), "IoT-Based Applications in Healthcare Devices," *J. Healthc. Eng.*, vol. 2021.

34. S. Siddiqui, M. Z. Shakir, A. A. Khan, and I. Dey, (2021), "Internet of Things (IoT) Enabled Architecture for Social Distancing During Pandemic," *Front. Commun. Networks*, vol. 2, p. 6.

35. I. de Morais Barroca Filho, G. Aquino, R. S. Malaquias, G. Girão, and S. R. M. Melo, (2021), "An IoT-Based Healthcare Platform for Patients in ICU Beds During the COVID-19 Outbreak," *IEEE Access*, vol. 9, pp. 27262–27277.

36. A. Akinola, G. Singh, I. Hashimu, T. Prabhat, and U. Nissanov, (2021), "FSS superstrate antenna for satellite cynosure on IoT to combat COVID-19 pandemic," *Sensors Int.*, vol. 2, p. 100090.

37. M. Nasajpour, S. Pouriyeh, R. M. Parizi, M. Dorodchi, M. Valero, and H. R. Arabnia, (2020), "Internet of Things for current COVID-19 and future pandemics: An exploratory study," *J. Healthc. informatics Res.*, pp. 1–40.

38. H. Mukhtar, S. Rubaiee, M. Krichen, and R. Alroobaea, (2021), "An IoT Framework for Screening of COVID-19 Using Real-Time Data from Wearable Sensors," *Int. J. Environ. Res. Public Health*, vol. 18, no. 8, p. 4022.

39. S. J. Alsunaidi *et al.*, (2021), "Applications of Big Data Analytics to Control COVID-19 Pandemic," *Sensors*, vol. 21, no. 7, p. 2282.

40. A. P. Ramallo-González, A. González-Vidal, and A. F. Skarmeta, (2021), "CIoTVID: Towards an Open IoT-Platform for Infective Pandemic Diseases such as COVID-19," *Sensors*, vol. 21, no. 2, p. 484.

41. S. Fahrni, C. Jansen, M. John, T. Kasah, B. Körber, and N. Mohr, (2020), "Coronavirus: Industrial IoT in challenging times," *McKinsey Co.*

42. R. P. Singh, M. Javaid, A. Haleem, and R. Suman, (2020), "Internet of things (IoT) applications to fight against COVID-19 pandemic," *Diabetes Metab. Syndr. Clin. Res. Rev.*, vol. 14, no. 4, pp. 521–524.

43. S. Malliga, S. V Kogilavani, and P. S. Nandhini, (2021), "A Comprehensive Review of Applications of Internet of Things for Covid-19 Pandemic," in *IOP Conference Series: Materials Science and Engineering*, vol. 1055, no. 1, p. 12083.

44. S. S. Vedaei *et al.*, (2020), "COVID-SAFE: an IoT-based system for automated health monitoring and surveillance in post-pandemic life," *IEEE Access*, vol. 8, pp. 188538–188551.

45. E. Elbasi, A. E. Topcu, and S. Mathew, (2021), "Prediction of COVID-19 Risk in Public Areas Using IoT and Machine Learning," *Electronics*, vol. 10, no. 14, p. 1677.

46. W. He, Z. J. Zhang, and W. Li, (2021), "Information technology solutions, challenges, and suggestions for tackling the COVID-19 pandemic," *Int. J. Inf. Manage.*, vol. 57, p. 102287.

47. J. Budd *et al.*, (2020), "Digital technologies in the public-health response to COVID-19," *Nat. Med.*, vol. 26, no. 8, pp. 1183–1192.

48. M. Umair, M. A. Cheema, O. Cheema, H. Li, and H. Lu, (2021), "Impact of COVID-19 on IoT Adoption in Healthcare, Smart Homes, Smart Buildings, Smart Cities, Transportation and Industrial IoT," *Sensors*, vol. 21, no. 11, p. 3838.

49. M. Ndiaye, S. S. Oyewobi, A. M. Abu-Mahfouz, G. P. Hancke, A. M. Kurien, and K. Djouani, (2020), "IoT in the wake of COVID-19: A survey on contributions, challenges and evolution," *IEEE Access*, vol. 8, pp. 186821–186839.

50. N. Sharma *et al.*, (2021), "A smart ontology-based IoT framework for remote patient monitoring," *Biomed. Signal Process. Control*, vol. 68, p. 102717.

51. C. Singh, (2019), "What is the Future Scope of IoT in Healthcare?", [Online]. Available: https://www.appventurez.com/blog/iot-healthcare-future-scope/.

CHAPTER 4

HIERARCHICAL ROUTING PROTOCOL FOR EVENT-BASED APPLICATIONS IN WIRELESS SENSOR NETWORKS

Rohit Srivastava
School of Computer Science, University of Petroleum and Energy Studies, Dehradun, India

In wireless sensor networks (WSN), the low-cost sensor nodes have limited computation and communication. Due to this, the amount of data transmission in the nodes needs to be restricted to increase the average lifetime of the node and to improve the overall bandwidth. In network, data aggregation is an effective approach inherently used for reduction in data transmission and energy consumed, thus, prolonging the network lifetime. In this work, a hierarchical routing protocol with in-network aggregation is proposed that has the objective to reduce the number of packet transmissions, remove redundancy, achieve reliable routing and high aggregation rate, etc. In this, we built a tree from sink to all the nodes in-network with the shortest path and then we aggregate the data based on the build route by intermediate nodes. Additionally, a failure detection mechanism is also introduced to perform accurate data delivery. The proposed Hierarchical based routing protocol is compared with Leach, and more system lifetime is achieved in the proposed algorithm which verifies the said approach.

4.1 INTRODUCTION

Networks and data play an important role in our day-to-day life, without which no work or application is possible. As in a wireless network (Nikolaos A. Pantazis, 2012), routing data efficiently towards the destination is the key aspect. The wireless network is a very wide field for research work and much of this research is proposed and now implemented in the real world. Many different techniques and routing algorithms (Subhajit Das, 2012) have been proposed for transferring data from the sensing nodes to the base station (BS). Many metrics such as packet delivery rate, efficiency, data aggregation, routing cost, loss of data, etc., are taken into consideration in evaluating any routing algorithm. All these routing algorithms (Nakas, 2020) have a focus on any of these metrics in their work but their main aim is to transmit data efficiently from different sensing nodes to sink (M. J. Islam, 2007) or BS. However, all have some advantages and disadvantages, as if we improve one metric then another metric gets disturbed.

Wireless sensor networks (WSN) (Rhim, 2018) are a class of remote organizations in which sensor hubs gather, measure and impart information obtained from the physical climate to an outside BS, consequently taking into consideration, observing and controlling the different actual boundaries. Information collection and in-network preparing procedures have been researched as proficient ways to deal with critical energy saving in WSNs (Ghaffari, 2014) by joining information from various sensor hubs at some collection reroute, thus, killing excess and limiting the quantity of transmission previously sending information to BS.

At first, in-network strategies (Pasquino, 2021) included diverse approaches to course bundles to join information coming from various sources yet coordinated towards the same destination(s). As such, these conventions were just directing calculations. In today's scenario, numerous extra examinations have been distributed, tending to the directing issue as well as instruments to address and consolidate information productively. In-network information total is a current issue that includes many layers of the convention stack and various parts of the convention plan, and a portrayal and characterization of ideas and calculations are ailing in the writing.

4.1.1 Problem Statement

The main problem with the sensor network is the collection of redundancy data. Sometimes, the same data is also produced by different sensor nodes. This will increase the number of messages having the same data that is to be transferred at sink or BS. Thus, due to this, communication costs got an increase in transferring those data to sink or BS, and then they have a load to remove redundancy data. Even, energy-constrained is also there, as sensor nodes have low power, so energy should be conserved to increase network lifetime. The main aim is to perform reliable routing with in-network data aggregation.

The proposed scheme will improve the energy level and reliability of the network while maintaining the good performance of the data routing through a wireless sensor network. This work is carried out to increase the lifetime of the network. This algorithm is event-based that is whenever simultaneous events are generated, redundancy data is removed and only one packet is sending forward. The data is sent hop-by-hop towards the sink so less energy is consumed. The route generated by the nodes has the largest aggregation points to remove redundancy.

4.2 LITERATURE SURVEY

According to (Elena Fasolo, 2007), in common sensor network situations, information is gathered by sensor hubs all through some space and should be made accessible at some focal sink node(s), where it is prepared, examined and utilized by the application. In-network conglomeration is the worldwide interaction of social affairs also, steering data through a multi-bounce organization, preparing information at moderate hubs to lessen asset utilization. For example, decreasing the large measure of force and data transfer capacity needed to handle the client's inquiry, in this way expanding network lifetime.

In information-gathering-based applications (Murukesan Loganathan, 2017), a significant number of correspondence parcels can be diminished by in-network collection, bringing about a more extended network lifetime. In-network collection has 2 methodologies that are In-network total with size decrease and without decrease. Fundamental pieces of In-network accumulation are steering convention, conglomeration capacity, and information portrayal. From this paper, it is identified as close itemized clarification of in-network accumulation and unique directing conventions are clarify in a nutshell. In this, they clarify their sorts; benefits, properties, and fundamental parts.

According to (Nikolidakis, 2013), Periodic collection conventions for information conglomeration are characterized into three gatherings for example intermittent basic, occasional per-bounce and intermittent per-jump changed. The foreordained measure of time totals all information got and afterwards advances the information at the host hub. Such a calculation is easy to carry out, however doesn't ensure the exactness of the information.

Occasional per-jump accumulation implies that every hub delays until it gets information from all kids, totals the information (Srivastava, 2001), and afterwards advances it toward the host hub. This methodology requires the utilization of a break if a portion of the kids does not react to the inquiry. Intermittent per-bounce changed is as the per-jump approach, aside from the break depends on the hub's situation in the directing tree. Hubs lower in the directing tree should encounter a break before hubs nearer to the host. This sort of falling break causes an "information wave" to spread up the tree toward the host (Abidoye, 2021) at standard stretches.

In the paper (Obraczka, 2004), the author investigates the issue of timing while amassing information in Remote Sensor Networks. They appear through numerical examination and reproduction that energy is saved by diminishing the measure of information moves in WSNs. They additionally clarify briefly about total period about information newness and energy saving.

As per the paper (Chen, 2013) a directing convention is a tree-based convention construct a customary most limited way directing tree. For example, the Shortest Path Tree (SPT) calculation utilizes an extremely straightforward technique to construct a directing tree in a conveyed design. In this methodology, each hub that recognizes an occasion reports its gathered data by utilizing the least way to the sink hub. Data combination happens at whatever point ways crossover (crafty data combination). In the paper, (Hong Luo, 2006) the ideal and imperfect accumulation of information accumulation is clarified into which three imperfect plans are clarified. SPT is one of these plans where steering tree and information course through least way calculation as it were.

4.3 DESIGN METHODOLOGY

In the proposed scheme, there will be four phases, which are as follows:

- Building Hop Tree and Parental Candidate Information Table Generation

- Event Generation and Leader Election
- Route Established by Leader and Update Hop Tree
- Data Transmission and Failure Detection

Phase I - Building Hop Tree and Parental Candidate Information Table Generation

The significant exercises in this stage are building the least order bounce tree set-up, parental up-and-comer data table, and directing development for every hub. HBIA structure progressive relations with Hop Construction Packet (HCP) (S. Prithi, 2020), which permits hubs to structure independent connections with no unified control. The HCP message contains 3 fields: ID, Hop_count, and Sink_id. The ID is the hub identifier that began or retransmitted the HCP message. The Hop_count field is the number of jumps from the sink hub. Sink_id shows which sink communicates the HCP bundle in the event of a different sink.

Source_Node	Hop_Count	Sink_Id
Hop Construction Packet		
Hop_Count	Parent_ID	Sink_Id

Parental Candidate Information

Figure 4.1 HCP packet format (S. Prithi, 2020).

The sink begins with bounce esteem "0", while other sensor hubs are "∞". At first, sink hub (S) or BS increment the Hop_count field by one and broadcast the HCP message < S1, 1, S1> to find one jump hub for example the sink broadcast to its neighbor hubs. Every hub hangs tight for a brief timeframe (T) to acquire at least one applicant guardian and records them in its Parental Candidate Information Table (PCIT). Every hub, after getting the HCP message, confirms if the worth of Hop_count in HCP message is not exactly the worth of Hop_count that it has put away, then, at that point, it keeps the bundle during T e.g., the worth of Hop_count field is 1, which is not exactly the bounce esteem ∞ of hub 1, and in any case, drops the bundle. On the off chance that the hour of T is done, hub 1 starts to select the parcels with the most minimal Hop Count esteems as its competitor guardians, and records the bundle data into PCIT. Hub 1 then, at that point raises the Hop_count field of the HCP bundle by 1 and rebroadcasts. Hub 6 also gets two-layer bundles from hubs 1 and 4 with a similar Hop_count field esteem. Consequently, the applicant guardians are the two hubs 1 and 4. Moreover, hub 1 gets an HCP parcel from hub 4, however, the jump worth of hub 1 is less than the Hop_count field of HCP message from hub 4. Thus, hub 1 overlooks the HCP parcel. Figure 4.2 shows the activity stream when a hub gets a parcel. Each hub keeps flooding the HCP parcel until the organization level is developed.

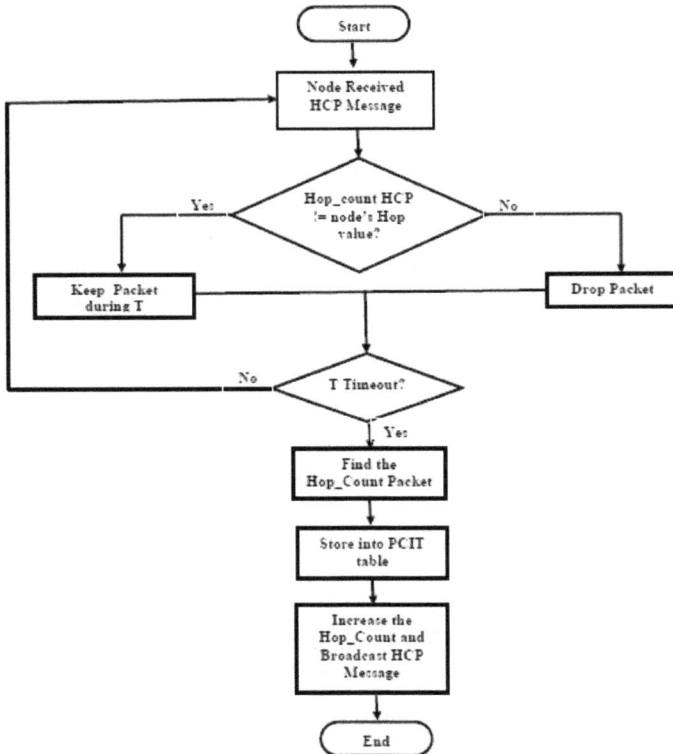

Figure 4.2 Flow chart for building Hop Tree.

Before the main occasion happens, there is no settled course, and the Hop_count variable stores the littlest distance to the sink. On the main occasion event, Hop_count will in any case be the littlest distance; notwithstanding, another course will be set up. After the main occasion, the Hop_count stores the more modest of two qualities: the distance to the sink or the distance to the nearest effectively settled course.

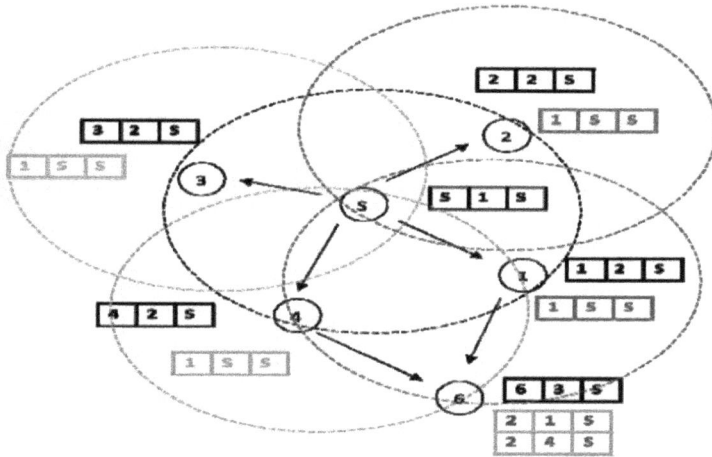

(a)

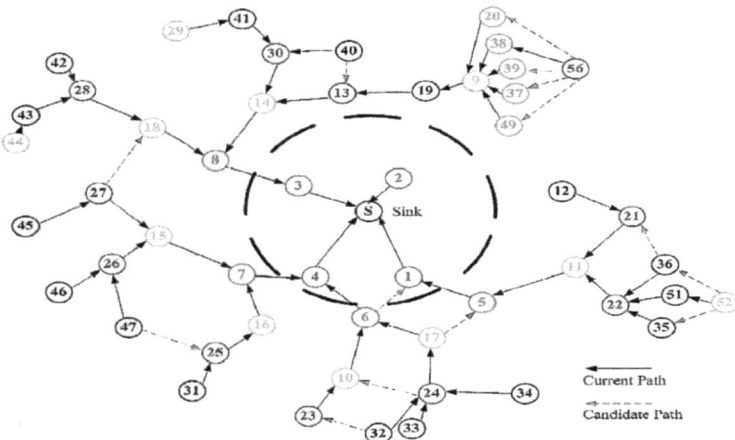

(b)

Figure 4.3 Building of Hop Tree (S. Prithi, 2020).

Phase II - Event Generation and Leader Election Algorithm

When one or more nodes detect the event, the leader election algorithm starts, and all sensing nodes are eligible for this election. We can say it cluster form, having all the sensing nodes who detect the event. If this is, the first event when the energy level of all the nodes is the same then the leader will be the node that is closest to the sink node. Otherwise, we have to check some conditions for leader election. Choose nodes with minimum HTT value and compare the residual energy of those nodes. The node having the highest energy level will become the leader i.e. Coordinator. Thus, the remaining nodes become Collaborators. The main work of the Coordinator is to aggregate the data that is sent by Collaborators and send them to the sink.

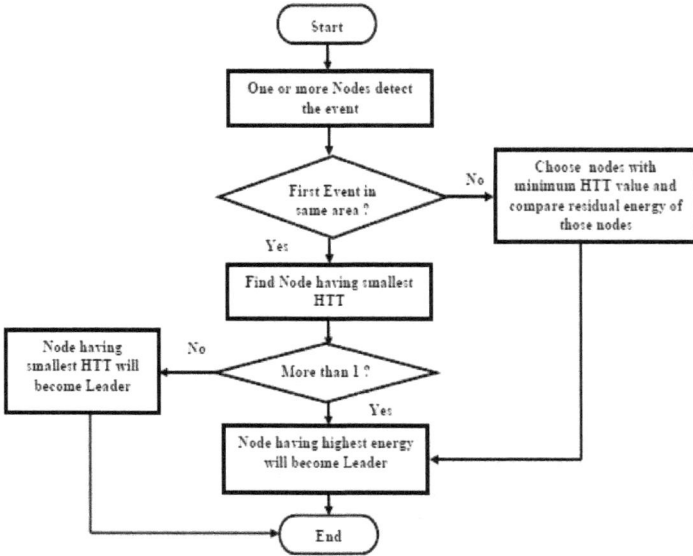

Figure 4.4 Flowchart of Leader Election.

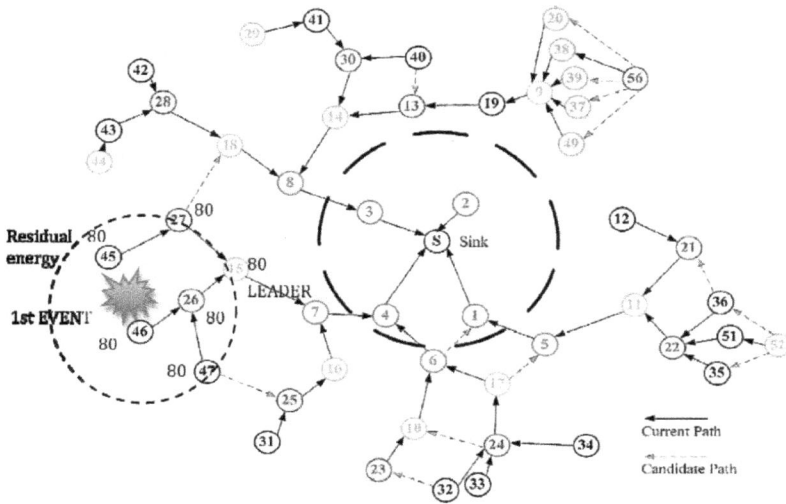

(a)

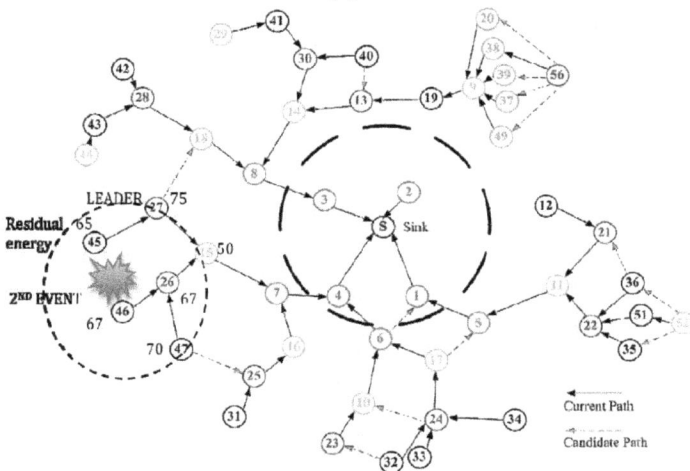

(b)

Figure 4.5 Leader Election Algorithm [16].

Phase III – Route Established and Update Hop Tree

Outside the cluster, the route is established by the leader (Coordinator) to the sink via the parent node. The Coordinator sends REM Route Established Message to its NextHop node i.e. parent node.

When a NextHop node receives the REM it will retransmit the REM to its NextHop and so on till sink reached or node that is part of the previous route.

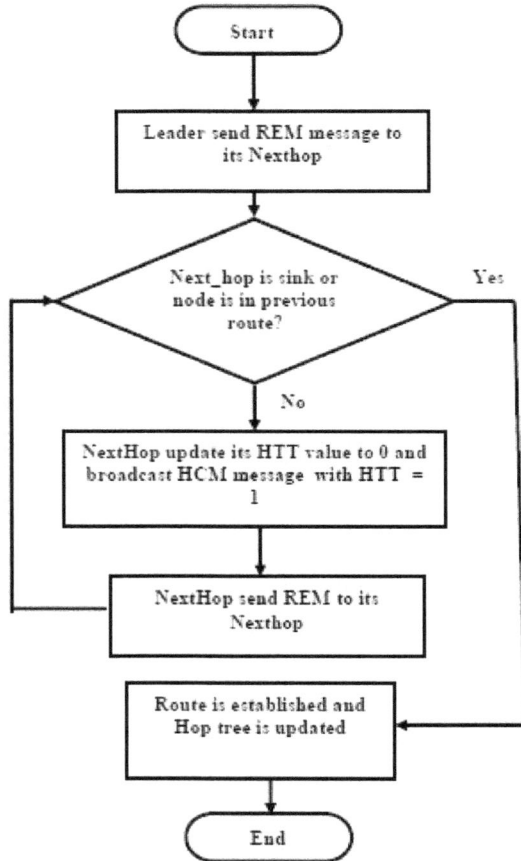

Figure 4.6 Flow chart of Route established and update Hop Tree.

The tree updating process starts in Relay nodes i.e. the nodes involved in the route established, by temporarily changing the HTT value. The Relay nodes initialized their HTT value to zero and broadcast HCM message with HTT value increment by 1 and same way all other nodes other than Relay nodes will continue the process. The main task of updating the tree is to consider the newly established route and also due to this we can increase the aggregation points.

Phase IV – Data Transmission and Failure Detection

Data is a transfer by the node to its ancestors i.e. NextHop. Here, a node has to check if there is more than one child then wait for some time and collect all the data from its child and send it to its NextHop. The Waiting Period for an aggregator node will be the average distance of collaborators nodes within the cluster and if it is outside the cluster then the average distance of event coordinators. Here, the average distance will be the number of hops distance between aggregator nodes and collaborator nodes. In this, we are trying to aggregate data at 3 levels: (i) At Collaborators Level (ii) At Coordinators Level (iii) Intermediate nodes outside the cluster. Due to this, a considerable number of packets are reduced, resulting in a longer network lifetime.

When a data packet is transferred by a node to its NextHop, then NextHop replies with Received Data Acknowledge (RDACK) to notice the source node that the data is successfully transmitted. On receive the RDACK from the NextHop, it can infer that it is alive. If it does not receive RDACK within a predetermined time, it is considered dead and selects the new NextHop having the highest energy.

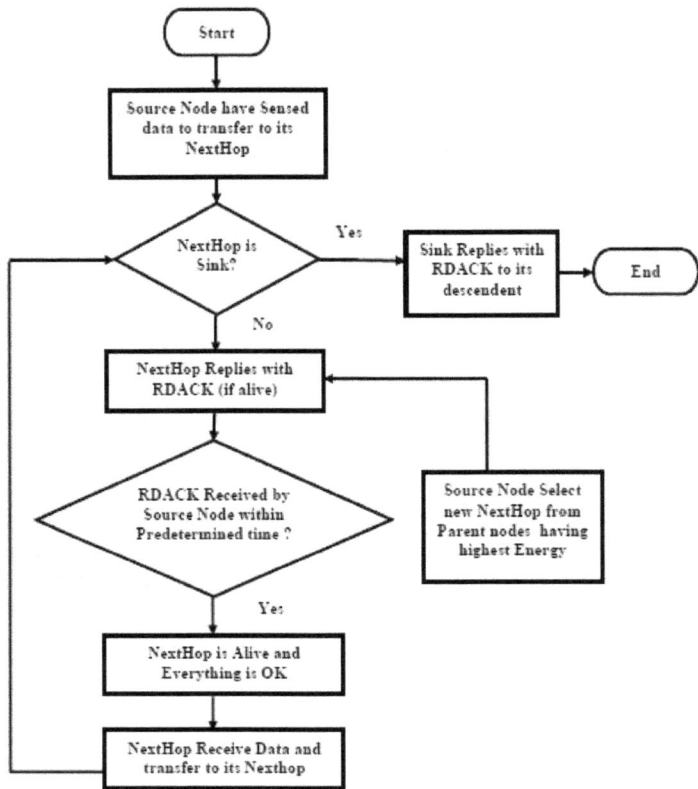

Figure 4.7 Flow chart of Data Transmission and Failure Detection.

4.4 IMPLEMENTATION AND RESULTS

The performance evaluation has been implemented in Matlab7. The following implementation parameters have been used:

- Network: The sensor nodes are randomly distributed over a sensor network field of (1000*1000) m2. Initially, a routing tree is generated using the shortest path towards the sink by using the communication radius (150m) of each sensor node.
- Event: This is an event-based routing protocol so initially 6 events are generated. Events are generated randomly, uniformly distributed, and occur at a random position. The event detection radius is (60m) that is the sensor nodes,

which come in this range, will detect the event. Initially, event-generating time is defined statically but afterwards, it will be done randomly.

- Energy: Initially, energy is allocated to each sensor node, except the sink node. The energy is utilized during routing the data and it depends upon the number of packets. The processing energy is the energy used in sorting out the data from the different packet that is received and make a single packet, which is forwarded to the parent node.

Here, the simulation is done in Matlab so m file, which performs, proposed hierarchical routing.

Hierarchical-routing Process

1. Defining the Parameters
2. Create Random Wireless Network
3. Generate Routing Table
4. Set Timer for event
5. Create Random Events at different area
6. Check which nodes detect the event-form cluster
7. Leader election algorithm initiated
8. The leader establishes a route towards the sink
9. Data is aggregated within-cluster and on route
10. The energy of every node is calculated
11. Check and calculate the alive nodes in every event
12. The algorithm stops when the maximum number of nodes dead
13. Generate a graph showing the lifetime of the network

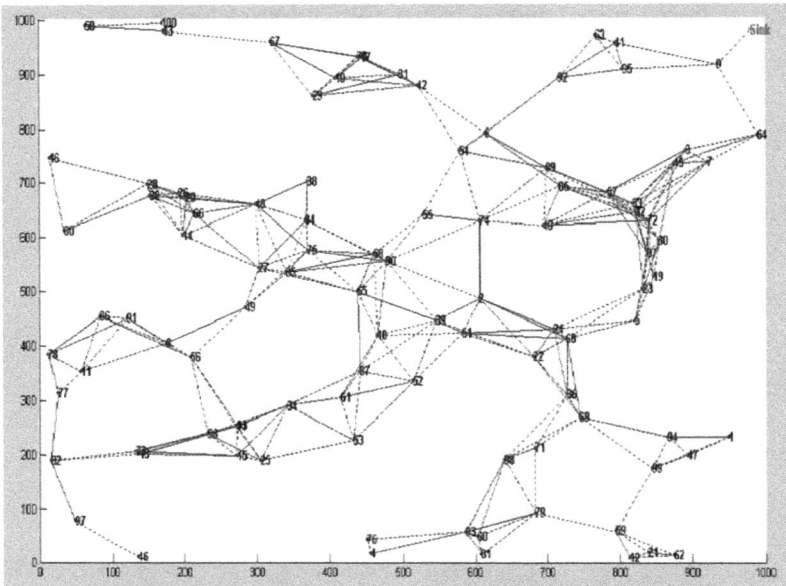

Figure 4.8 Building Routing Tree.

Figure 4.8 shows the wireless network which is built randomly, a red color node is a sink that is located right-top in the network. The node is black color numbered and the blue color link connects the nodes which come in its radius of communication range. The network area is 1000*1000 m² in which the 100 nodes are distributed randomly.

Table 4.1 Routing Table of Complete Network

No.	Hop Distance	Node	Parent/Neighbor
1	10	1	47
2	9	1	94
3	9	1	99
4	8	2	22
5	7	2	24
6	8	2	39
7	8	2	51
8	7	2	66
9	7	2	74
10	8	2	90
11	4	3	7

12	4	3	32
13	4	3	33
14	4	3	45
15	4	3	57
16	3	3	64
17	4	3	72
18	12	4	76
19	10	4	83
20	6	5	19
21	6	5	23
22	7	5	24
23	6	5	37
24	7	5	68
25	5	6	42
26	5	6	54
27	5	6	85
28	5	6	89
29	4	6	92
30	4	7	3
31	4	7	32
32	4	7	33
33	4	7	45
34	3	7	64
35	4	7	72
36	11	8	11
37	10	8	49

Table 4.1 shows the routing table generated according to Hop distance from the sink and also gets information about the parent node and neighboring nodes. With neighboring nodes, they also calculated the hop distance from the sink. Initially, every node in the network will consider the shortest path which is less hop distance. After that, in case of failure or less energy, it will consider neighboring nodes for transferring data.

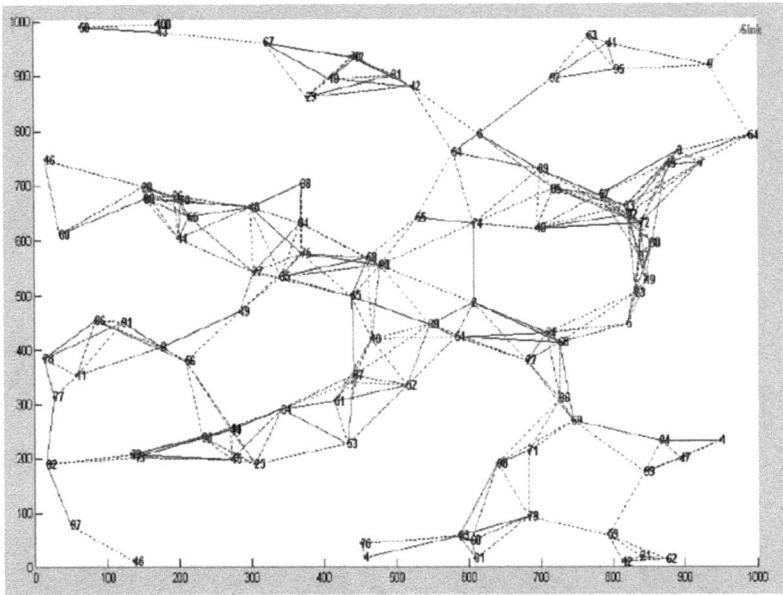

Figure 4.9 First Event Generation.

In Figure 4.9, 1ˢᵗ event is generated with a red dot and red lines are the linked nodes with had detected those nodes. After a particular time interval, other events are also generated and similarly, the nodes, which are in that range of event, will detect the event. Here, only 6 events are shown, which are going to hold in a time interval that has been decided.

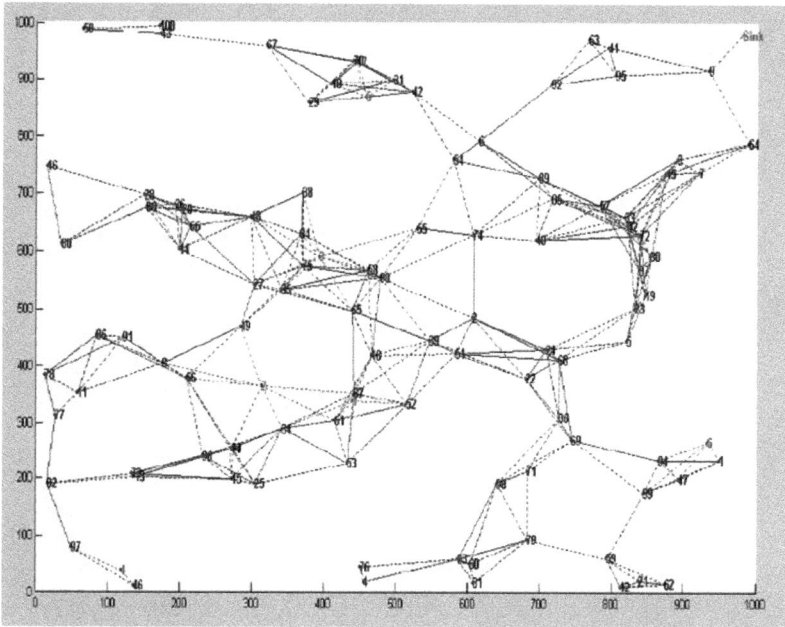

Figure 4.10 All Six-Event Generation.

In Figure 4.10, all 6 events are generated with delay time and also link the nodes, which detect that event.

Table 4.2 Routing Table for Event

No.	Event Nos.	Detect Node	Min Hop
1	1	18	8
2	1	34	10
3	1	52	8
4	1	53	9
5	1	61	9
6	1	67	9
7	2	8	10
8	2	34	10
9	2	44	11
10	2	49	9
11	2	56	10
12	2	61	9
13	2	87	9
14	2	93	11

15	3	27	9
16	3	35	8
17	3	38	9
18	3	48	9
19	3	55	6
20	3	58	7
21	3	65	8
22	3	75	8
23	3	84	8
24	3	90	7
25	4	46	15
26	4	97	14
27	5	10	6
28	5	17	6
29	5	29	6
30	5	31	6
31	5	42	5
32	5	70	6
33	6	1	9
34	6	47	9
35	6	94	8

Table 4.2 presents the list of nodes, which have detected a particular event at a particular time. This node now going to form a cluster of that event and then out of which a leader is selected based on residual energy. Initially, each node has the same energy so the node with minimum hop distance is selected as the leader.

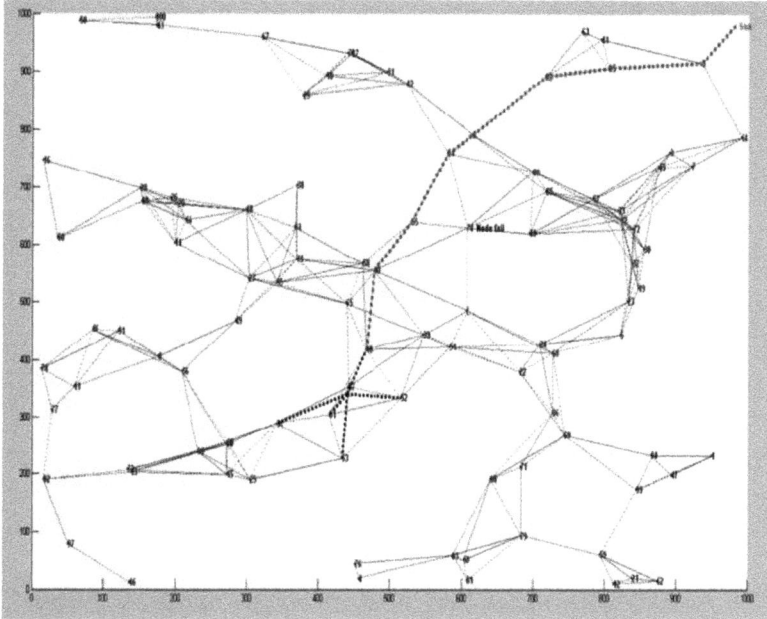

Figure 4.11 Route Change when Node Failure Detection.

In Figure 4.11, initially, the route is routed via the 74th node but when the route is established it will check the energy level of the node. In this, the energy level of the 74th node is zero so the route will be generated through another node i.e. 90th node.

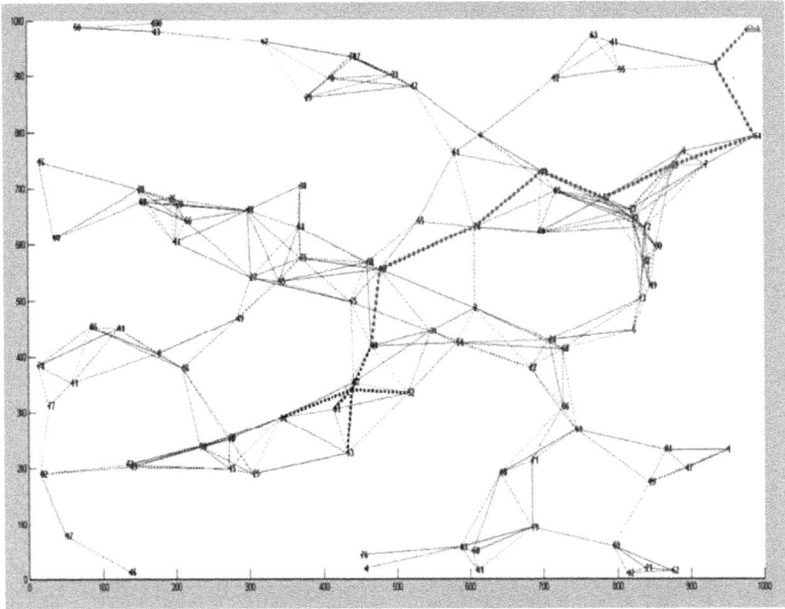

Figure 4.12 Route Established by Event.

In Figure 4.12, the leader establishes the route. The red color shows the path generated towards the sink. The data is gathered and sent via this route.

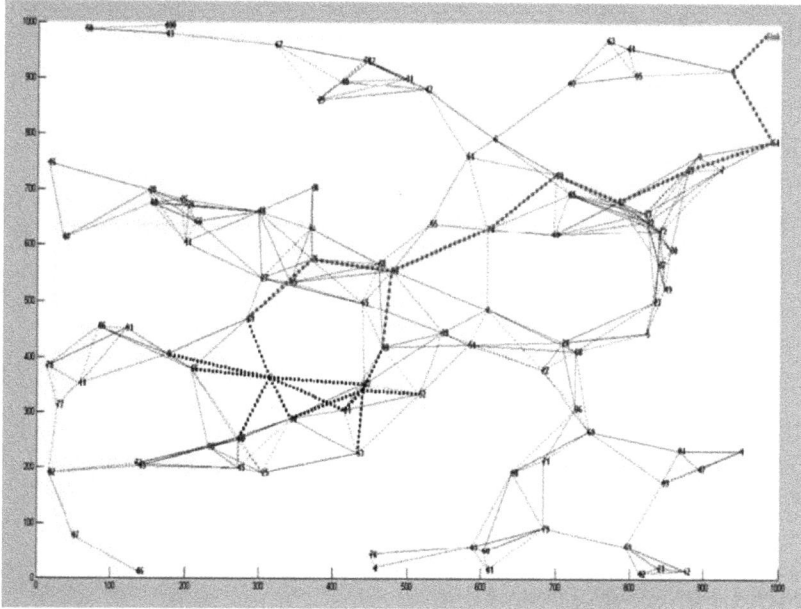

Figure 4.13 Routes Overlap and Data Aggregated at an Intermediate Node.

In Figure 4.13, the routes overlap, which is generated by 2 events, and thus data get aggregated at an intermediate node, which is outside the cluster. Thus, redundancy reduces at this level. The intermediate node merges the data and forwards it. An energy-efficient hierarchical-based routing protocol is developed; it has also worked for data aggregation as this use all 3 types of data aggregation technique. Using energy concepts at each level of data sending makes an energy-efficient routing algorithm. The simulation uses the same network parameter. The simulation runs for a maximum of 5000 rounds and checked the number of alive nodes with time. Table 4.3 shows the number of alive nodes for the time intervals.

Table 4.3 Number of Alive Nodes for Time Intervals

Number of Rounds	Leach	Proposed Algorithm
0	100	100
1000	100	100
1500	90	94
2000	50	65
2500	15	30

3000	4	15
3500	0	4
4000	0	0

Network Lifetime

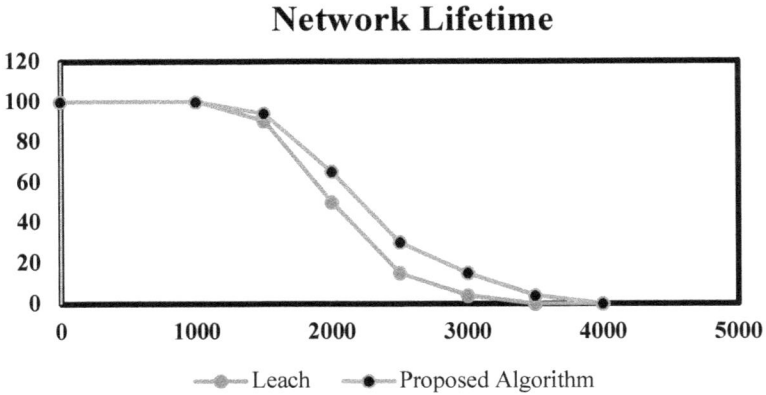

—●—Leach   —●—Proposed Algorithm

Figure 4.14 Network Lifetime Graph.

From Table 4.3 it can be seen that different number of nodes alive as time passed. However, the time when the nodes were exhausted was less in the proposed algorithm rather than in Leach. Therefore, the network lifetime was increased almost for some time using the proposed algorithm. Figure 4.14 shows the difference of time between Leach and the proposed algorithm when the nodes were exhausted. Therefore, the proposed scheme for energy efficiency has improved the routing algorithm. As the proposed algorithm is compared with energy-efficient Leach and this simulation result shows that the proposed algorithm is better than Leach to some extent. This increases the lifetime of the network.

4.5 CONCLUSION AND FUTURE WORK

A brief study on wireless sensor networks and their routing protocols is done. All the routing protocols introduced until now have their pros and cons. One important aspect of WSNs is data in-network aggregation. In event-based applications of WSNs, aggregation aware routing algorithm plays an important role. Here, a hierarchical-based routing protocol is proposed which routes the data through a tree structure network and maximum aggregation is performed by increasing aggregator nodes in the data routing. This will remove redundancy in the data that is sent by sensor nodes and also reduce the number of messages that are to be sending towards BS. In addition, energy is considered as a constraint in our protocol to make it an energy-efficient algorithm and also offering a fault-tolerant mechanism to improve data delivery rate. A waiting time at the aggregation node is introduced which is based on the distance that is the number of hops of the event coordinators.

In future work, control messages should be taken into consideration. In this, it requires more control messages for all this process so it can be minimized. In addition, all the nodes are assumed as stationary so new routing strategies will be devised for dynamic sensor nodes and also the construction of a routing tree that meets application needs.

References

1. Abidoye, A. K. (2021). Energy-efficient hierarchical routing in wireless sensor networks based on fog computing. J Wireless Com Network, 8(1), 1687-1499.
2. Chen, S. W. (2013). LCM: A Link-Aware Clustering Mechanism for Energy-Efficient Routing in Wireless Sensor Networks. IEEE Sensors Journal, 13(2), 728-736.
3. Elena Fasolo, M. R. (2007). In-network Aggregation Techniques for Wireless Sensor Networks: A Survey. IEEE Communications Surveys, 1(4), 1-26.
4. Ghaffari, A. (2014). An Energy-Efficient Routing Protocol for Wireless Sensor Networks using A-star Algorithm. Journal of Applied Research and Technology. JART, 12(4), 815-822.
5. Hong Luo, J. L. (2006). Adaptive Data Fusion for Energy Efficient Routing in Wireless Sensor Networks. IEEE Transactions on Computers, 55(10), 1286-1299.
6. M. J. Islam, M. M. (2007). A-sLEACH: An Advanced Solar Aware Leach Protocol for Energy Efficient Routing in Wireless Sensor Networks. Sixth International Conference on Networking (ICN'07). Shanghai.
7. Murukesan Loganathan, T. S. (2017). Energy-efficient routing protocols for wireless sensor networks: comparison and future directions. International Conference on Emerging Electronic Solutions for IoT. Penang, Malaysia.
8. Nakas, C. a. (2020). Energy-Efficient Routing in Wireless Sensor Networks: A Comprehensive Survey. Algorithms, 13(3), 72.
9. Nikolaos A. Pantazis, S. A. (2012). Energy-Efficient Routing Protocols in Wireless Sensor Networks: A Survey. IEEE Communications

Surveys, 1-41.

10. Nikolidakis, S. A. (2013). Energy-Efficient Routing in Wireless Sensor Networks Through Balanced Clustering. Algorithms, 6(1), 29-42.

11. Obraczka, I. S. (2004). The impact of timing in data aggregation for sensor networks. IEEE International Conference in Communications. Malaysia.

12. Pasquino, N. Y. (2021). Energy-Efficient Routing Protocol Based on Zone for Heterogeneous Wireless Sensor Networks. Journal of Electrical and Computer Engineering, 10(11), 5557756.

13. Rhim, H. T. (2018). A multi-hop graph-based approach for an energy-efficient routing protocol in wireless sensor networks. Human-centric Computing and Information Sciences, 8(1).

14. S. Prithi, S. S. (2020). LD2FA-PSO: A novel Learning Dynamic Deterministic Finite Automata with PSO algorithm for secured energy-efficient routing in Wireless Sensor Network. Ad Hoc Networks, 97(1), 102024.

15. Srivastava, C. S. (2001). Energy-efficient routing in wireless sensor networks. MILCOM Proceedings Communications for Network-Centric Operations: Creating the Information Force. McLean, VA, USA.

16. Subhajit Das, S. B. (2012). Energy-Efficient Routing in Wireless Sensor Network. Procedia Technology, 6(1), 731-738.

CHAPTER 5

ARTIFICIAL NEURAL NETWORK BASED ANTENNA DESIGNS: A SELECTIVE REVIEW

Manpreet Kaur[1], Jagtar Singh Sivia[2] and Navneet Kaur[3]
[1,2]YCoE, Punjabi University Patiala South Campus, Talwandi Sabo, Punjab, India
[3]Punjabi University, Patiala, Punjab, India

Artificial neural network (ANN) is defined as an efficient data computational model that is inspired by the biological neural network to perform various complex tasks in an easy way. ANN approach appears to be more appropriate for designing high precision antennas. It has been successfully applied in various domains. ANNs have been receiving great popularity and attention from researchers belonging to different streams. This paper deals exclusively with the application of the ANN technique for the analysis and synthesis of various existing antennas. Antenna optimization aims at producing advanced electromagnetic structures that show superior performance. The design of an antenna mainly depends on the targeted application and the operational frequency range. Training a network means weight adjustments of the neurons using a suitable learning algorithm. The analysis is essential to evaluate fundamental qualities and drawbacks associated with the antenna. The selection of the optimization approach is highly linked with the nature of the antenna design problem and the designer's experience. Several antennas designs using this strategy are studied and compared. It briefly highlights the role of neural networks in the estimation of fundamental performance parameters. The purpose of this literature survey is to understand and analyze the process of antenna designing using ANNs. The design specifications and electromagnetic performances have been taken into account. All sorts of antennas are based on the principle of electromagnetic theory. In all the antennas, due to its generalization capability, the ANN approach is preferred for estimating distinct antenna parameters. The understanding of important aspects concerned with the antennas allows us to investigate some preliminary characteristics. Additionally, a comparison among different ANN-based antennas is done to realize the feasibility of the suggested strategy.

5.1 INTRODUCTION

In the present era of communication, the mode of exchanging data is highly modernized. Everyone is highly dependent upon technology for daily requirements. Modern communication systems consider an antenna as the most important and indispensable component. An antenna is a specialized form of transducer whose purpose is to transmit or receive electromagnetic waves [22]. During transmission, it converts the alternating current into electromagnetic waves. Conversely, during the reception, it converts the electromagnetic waves back into an alternating current. The operation in these two modes is based on the reciprocity property. Usually, an antenna consists of an arrangement of a particular type of metallic conductors attached to the transmitter or receiver [24]. The antennas can be realized to effectively transmit electromagnetic waves equally in all directions or a particular direction Several types of antennas are available that differ in their physical structure, operational frequency range, and modes of operation [25]. Nowadays, various renowned scientists and researchers have been working hard for the development of efficient and miniaturized antennas [27].

5.2 MICROSTRIP PATCH ANTENNA

In 1953, G. A. Deschamps was the first person who introduced the concept of microstrip antenna. But this work had not become practical till the 1970s [22]. Afterwards, in 1972, the first practical microstrip antenna was invented by Robert E. Bob Munson. Microstrip patch antennas belong to an important class of planer antennas that have been extensively developed in the last few decades [28]. Designing such types of antennas has become a topic of immense interest for researchers because these antennas exhibit significant advantages such as small volume, low manufacturing cost, mechanically robustness, compatibility with microwave integrated circuits, and so on [7]. In these antennas, the commonly used frequency range is 1 to 100 GHz. The microstrip antenna is a dielectric substrate placed between two conducting layers. The upper conducting plate is called a patch whereas the lower conducting plate is termed as a ground plane [25]. The plates are very thin and are made up of either copper or gold. The commonly available shapes of the patch are rectangular, circular, elliptical, square, and so on. The length of the patch is based on the fundamental condition i.e. $\lambda_0 /3 < L < \lambda_0$ /2 [22]. The edges of the patch experience fringing effects due to finite dimensions along both sides. The fringing fields through the open sidewalls result in the phenomenon of radiation [26]. Substrate materials with dielectric constants $2.2 \leq \varepsilon_r \leq 12$ are commonly available. For good performance characteristics, the preferred thick substrate is having a low dielectric constant value. However, it leads to an enhancement in antenna size. For small-sized antennas, materials having high dielectric constant values are desirable. But, because of greater losses, these structures exhibit narrow bandwidth and are less efficient [24]. Therefore, there exists a contradiction between antenna size and good electromagnetic performance. These antennas have the strong capability of transmission and reception over long distances and are useful in several domains such as aircraft, radio-frequency identification systems, satellites, radars, missiles, and other wireless applications [25].

5.3 ARTIFICIAL NEURAL NETWORK

Artificial neural network (ANN) modeling is a promising technique used for solving non-linear problems related to science and engineering. ANN is a special type of biologically inspired classification network that follows the biological structure of the human brain [1]. For efficient functionality, modern technology employs artificial neural networks because these networks are extremely adaptive and have the capability to learn very quickly [2]. ANN has multiple neurons that are arranged in the form of layers. The first layer, i.e., input layer is fed with the inputs from the outside world. The intermediate layer, i.e., hidden layer performs the required computations and then forwards the output to the next layer [3]. At the last stage, the output layer generates the overall output. In brief, the size of the input layer is determined by the number of actual inputs, and the size of the output layer is associated with the number of outputs to be estimated [4]. The size associated with the hidden layer varies according to the user requirements and is determined by the hit and trial approach. During the training process, weights and threshold values are automatically adjusted by the network so that the computed error exhibits a minimum value [5]. Absolute error represents the difference between the actual and estimated values. The adjustments are carried out by the backpropagation mechanism. Figure 5.1 shows the basic structure of ANN [2].

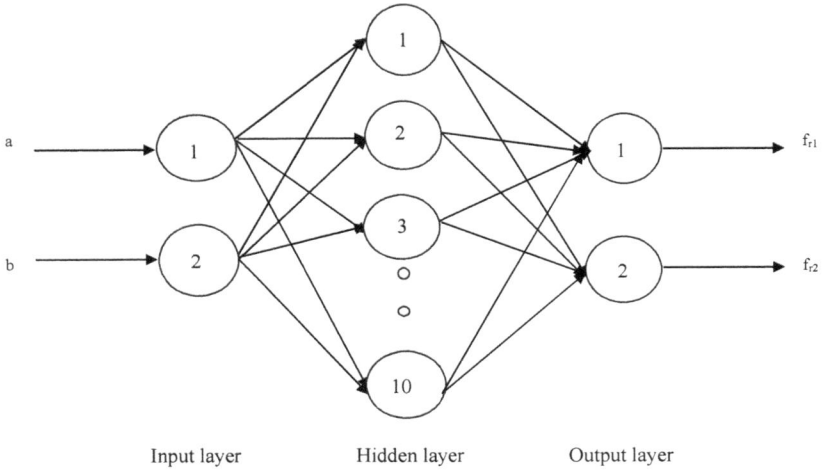

Figure 5.1 ANN structure [2].

5.4 SIGNIFICANCE OF ANN APPROACH IN ANTENNA DESIGNING

ANNs are considered efficient computational tools that collect information through experience, generalize from previous evaluations and then generate the relevant outcomes [2]. This concept is also utilized in electromagnetic theory and antenna designing. Antennas play an important role in communication purposes. The analysis and synthesis of any kind of antenna can be done with the ANN technique. For this process, a set of samples are to be created. From these samples, some samples are employed for training and the remaining samples are for testing and validation. The error is assessed by using the following expression [5].

Absolute error = [Simulated outcome − ANN model outcome]   (1)

This chapter presents a brief literature review on various antennas that are designed with ANN strategy. The research work of different authors is studied and reviewed. After that, important findings are highlighted. The entire paper is organized into seven sections. Description of microstrip antenna is Section 2. A brief overview of the artificial neural network approach is presented in Section 3. The significance of the ANN technique in antenna designing is given in Section 4. A comprehensive literature review on the different types of antennas is elaborated in Section 5. A comparison of few already available ANN-based antennas is done in Section 6. In the end, the Conclusion is drawn in Section 7.

5.4.1 Different ANN-based Antenna Designs

The design process of different types of antennas using ANN is described below:

Zooghby *et al.* (1999) introduced a unique approach for solving an important problem of adaptive beamforming. The weight values of the adaptive arrays were computed through the Radial Basis Function Neural Network (RBFNN) model. In satellite communication as well as in global positioning systems, signals continuously change their directions. Therefore, there is a need for a fast-tracking system to track the desired users and cancel the effect of the interfering signals. These arrays are responsible for increased system capacity for the existing communication systems Simulations of 1-D and 2-D arrays were done successfully [6].

Guney *et al.* (2004) described the resonant frequency estimation procedure of microstrip antennas having thin and thick substrates with the ANN technique. A measured data set was used to train the neural network. Resonant frequencies obtained with different algorithms employed for training were computed and then compared. As the neural network provides good accuracy and less computational effort, therefore it was suggested that this approach is appropriate for the development of fast CAD algorithms. The authors also illustrated that the proposed ANN approach can be feasible for engineering applications [7].

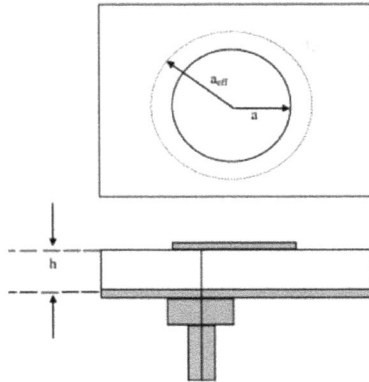

Figure 5.2 Circular microstrip antenna [8].

Ganguar *et al.* (2008) described the circular patch antenna based on the ANN strategy. The salient features of ANN include fast learning capability, ease of implementation and less consuming time. The three-layered model had 3-15-5 neurons at the respective layers. In this work, a back propagation algorithm was selected for training purposes. The considered values of the learning rate, momentum factor and normalized system error were 0.9, 0.5 and 0.00001 respectively. ANN computed resonant frequencies were similar to theoretical computed and measured values [8]. The structure of the circular microstrip antenna is represented in Figure 5.2 [8].

Thakare *et al.* (2009) used an ANN approach for enhancing the bandwidth of an inset-fed microstrip antenna designed to resonate at 10 GHz. The structure was modified by adding slots in the design. Slot dimensions were assumed as input to the neural network and output was examined in the form of bandwidth. The neurons fixed at the first, second and third layers were 4, 25 and 2 respectively. Two different algorithms were implemented, and their outcomes were compared with the simulated and practically observed outcomes [9].

Akdagli *et al.* (2013) reported the use of ANN for the examination of an operating frequency of 144 different E-shaped antennas. Simulated resonant frequencies of all antennas were determined with IE3D software by taking different values of design parameters. Antenna fabricated on Rogers RT/ duroid 5870 substrate provided resonance at 2.4 GHz with a gain of 6.88 dBi. The reported average percentage errors during training and testing were 0.257% and 0.523% respectively. Figure 5.3 illustrates the designed E-shaped antenna[10].

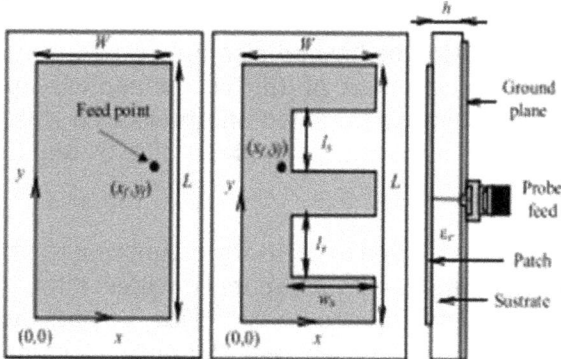

Figure 5.3 E-shaped antenna [10].

Can *et al.* (2013) develop an ANN-based PIN-loaded antenna. The multilayered structure had neurons associated with the input layer and two neurons associated with the output layer. Implemented antennas were evaluated with both FEM based and ANN solvers. The average error corresponding to lower frequency and upper frequency was 7.92% and 2.47% respectively. The effect of permittivity, antenna size and position of shorting pin was also evaluated [11]. Dhaliwal *et al.* (2013) reported the use of ANN for the analysis of the Sierpinski gasket fractal antenna. Three different ANN models were implemented and their results were compared. In each model, the input parameters assumed for the analysis were dielectric constant, thickness, side length and iteration order. The examined outcomes were the values of the resonant frequency. The behaviour was analyzed in terms of mean absolute error and the coefficient of correlation. It was found that RBFNN showed good performance in comparison to other models [12]. The designed antenna is shown in Figure 5.4 [12].

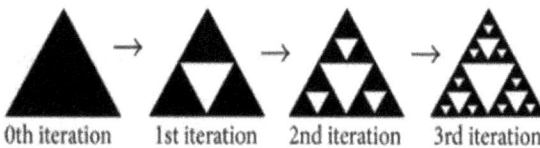

Figure 5.4 Sierpinski gasket antenna [12].

Sivia *et al.* (2013) analyzed the behaviour of a circular fractal antenna which was fabricated on Roger RT 5880 Duroid substrate. ANN approach was used for the performance evaluation of realized antenna. The fractal approach used in the design process resulted in multiband characteristics. The number of operating bands was examined by the iteration order and also linked with the shape of the geometry. The effectiveness of the recommended approach was proved by the estimated, simulated and measured results [13]. The implemented circular fractal antenna is illustrated in Figure 5.5 [13].

Sivia *et al.* (2013) employed an ANN for estimating the side length of the implemented antenna. Four inputs considered for the proposed model were resonant frequency, substrate height, dielectric constant and iteration number, whereas the output estimated was the side length of the design. 177 epochs were taken to provide the mean square error of 8.74e-007. The absolute error computed between the simulated and estimated value was 0.004515. The authors stated that close agreement among the estimated and simulated data strongly supported the validity associated with the hybrid model [14].

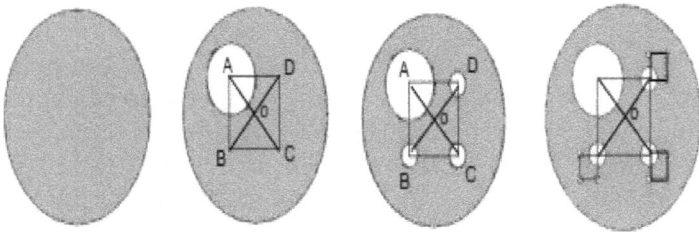

Figure 5.5 Implemented circular fractal antenna [13].

Aneesh *et al.* (2014) investigated the triple-band operation of slot loaded patch antenna with ANN. Patch antenna loaded with two parallel slots and notches was analyzed theoretically with circuit theory concept. From a data pool of 61 samples, RBFNN was trained and tested with 40 and 21 samples respectively. The notch loaded patch had four inputs and one output. It was examined that at 10.2 GHz, observed simulated and theoretical calculated values associated with antenna efficiency were 78% and 71%, whereas, at 17.2 GHz, these values were 73% and 70% [15]. The fabricated model is shown in Figure 5.6 [15].

Figure 5.6 The fabricated model [15].

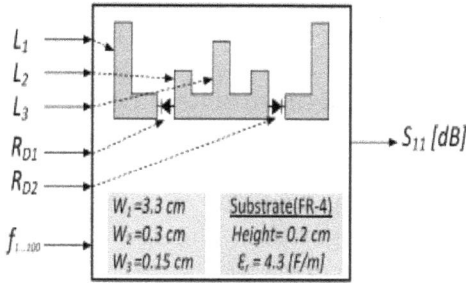

Figure 5.7 The implemented antenna [17].

Khan *et al.* (2015) demonstrated the operational performance of microstrip antennas. A comprehensive survey on the analysis and synthesis of few antennas with different neural networks was presented. Initially, a single performance parameter of different geometries was computed. Then, the ANN approach was employed for the simultaneous evaluation of different performance parameters. It was noted that the trained neural network model showed almost similar outcomes to that of simulated and measured ones [16].

Simsek *et al.* (2016) illustrated the modeling of a reconfigurable antenna by following a knowledge-based three-step procedure. ANN is an efficient modeling approach for solving engineering related problems based on the generated input/output data sets. The objective behind the modelling problem was to examine the S_{11} value corresponding to the frequency. Results mentioned that three-step modelling performs better than the conventional ANN approach. This method also reduces the time consumption associated with non-linear and complex problems [17]. The implemented antenna is shown in Figure 5.7 [17].

Gehani *et al.* (2017) worked on the utilization of the hybrid neuro-fuzzy technique for the development of the Sierpinski carpet fractal antenna. The hybrid neuro-fuzzy system showed faster and accurate learning because of its ability to integrate the fuzzy system and ANN. The model performed the analysis and synthesis in the range of 0.5 to 3 GHz and the effectiveness was assembled from the error value. The trained model resulted in return loss characteristics based on the geometrical parameters. It was concluded that considered membership functions provide almost similar results [18].

Singh *et al.* (2017) designed a modified circular patch antenna of radius 33 mm with an ANN approach. The benefit of using ANN is to tackle complex problems because of its superior computational ability. Operating frequencies were estimated by taking different dimensions of the antenna. It was demonstrated that the Levengberg-Marquardt (LM) algorithm provided fast and accurate results than the Quasi-Mewton BFG algorithm. The antenna provided triple frequency operation, thus covering UHF and C bands [19].

Khan *et al.* (2018) designed an E-shaped antenna with ANN-based on LM, Scaled Conjugate Gradient (SCG) and One Step Secant (OSS) algorithms. A miniaturized E-shaped antenna was created by incorporating two parallel slots in the rectangular structure. 144 samples with different design specifications were generated with IE3D software. The Average percentage errors (APE) value calculated with LM, SCG and OSS was 0.1689%, 0.988% and 1.229% respectively. The calculated values of APE with all three algorithms were compared and concluded that the best and accurate results were provided with the LM algorithm [20]. The suggested E-shaped antenna is shown in Figure 5.8 [20].

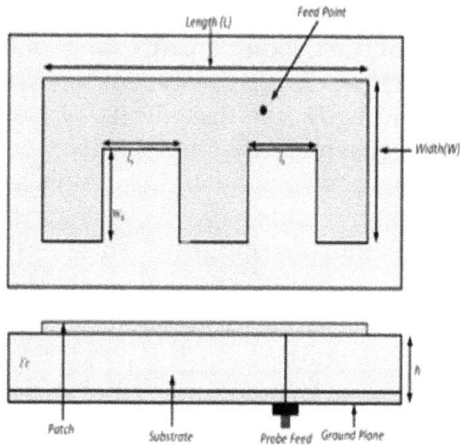

Figure 5.8 Suggested E-shaped antenna [20].

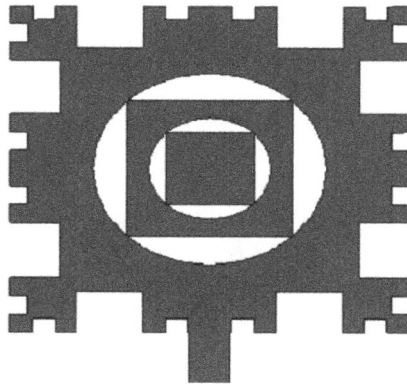

Figure 5.9 Hybrid fractal antenna design [2].

Kaur *et al.* (2019) developed an ANN model for compact dual-band hybrid fractal antenna that showed functionality at Wireless Medical Telemetry Service (WMTS) and Industrial, Scientific and Medical (ISM) bands. From practical observations, it was depicted that the resonant frequencies were 1.4290 GHz and 2.4380 GHz with S(1,1) values -12.4 dB and -15.72 dB respectively. It showed a gain of 3.63 dB and 7.80 dB at the claimed frequencies [2]. The hybrid fractal antenna design is shown in Figure 5.9 [2].

Dhaliwal *et al.* (2020) proposed a novel two-step method for the design optimization of fractal antenna. ANN modelling is considered an effective approach for design optimization. The two-step approach utilized in the work was based on ANN output sensitivity and optimization of ensemble members with a genetic algorithm. The proposed technique was validated by using it over six benchmark functions. Resultant geometry was verified through measurements and provided miniaturization of 48.34% [21]. Figure 5.10 represents the design process of fractal geometry [21].

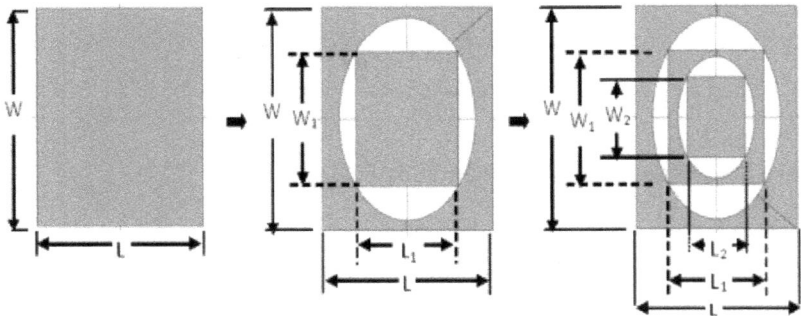

Figure 5.10 Design process of fractal geometry [21].

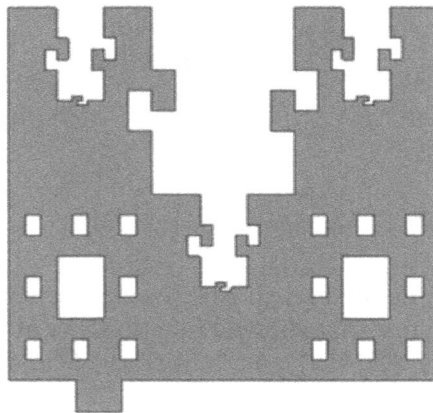

Figure 5.11 Realized hybrid fractal antenna [3].

Kaur *et al.* (2020) reported a compact Giuseppe Peano, Cantor Set and Sierpinski Carpet fractals-based hybrid fractal antenna is designed which is useful for ISM band applications. The hybrid fractal design was created by fusing Giuseppe Peano, Cantor set and Sierpinski carpet fractals together. The optimization of the microstrip line feed position was done by using a firefly algorithm. The proposed GCSA showed resonance at 2.4450 GHz with S(1,1) < -10 dB [3]. The realized antenna is shown in Figure 5.11 [3].

5.5 COMPARISON OF EXISTING ANTENNAS

Table 5.1 presents the comparative analysis of existing microstrip antennas.

Table 5.1 Comparative Analysis of Antennas

Ref. no.	Patch size (mm²)	Designed structure	Type of application
[1]	29 x 29	Integration of Minkowski, Giuseppe Peano and Koch curves	Biomedical
[2]	29 x 29	Combination of Minkowski and Circular-rectangular fractals	Biomedical
[3]	29 x 29	Fusion of Giuseppe Peano, Cantor Set and Sierpinski Carpet fractals	Biomedical
[4]	29 x 29	Fusion of Giuseppe Peano and Cantor set fractals	ISM (2.45 and 5.8 GHz)
[6]	--	Neural network approach for finding the weights of one- and two-dimensional adaptive arrays.	Communication systems
[7]	--	Neural models based on multi-layered perceptrons are described for the computation of resonant frequency of rectangular microstrip antennas with thin and thick substrates	--
[9]	6 x 8.88	ANN technique was used to design an inset feed microstrip antenna in which slots were introduced for bandwidth improvement.	Wideband
[10]	--	ANN was proposed to compute the resonant frequency of E-	--

		shaped compact microstrip antennas	
[11]	37.3 x 24.8	Estimation of the operating frequencies of shorting pin-loaded rectangular patch antennas using ANN	Wireless communication
[14]	50 x 50	ANN-based design of Sierpinski carpet fractal antenna.	Multiband
[15]	6 x 8	Parametric analysis of the antenna is done with the ANN model.	X- and Ku-bands
[17]	--	Three-step modeling strategy is applied to obtain S_{11} of antenna design parameters corresponding to the frequency.	--
[18]	77 x 77	Application of hybrid neuro-fuzzy model for the analysis and synthesis of square Sierpinski carpet fractal antenna	Multiband
[20]	25 x 20	E-shaped microstrip antenna using ANN	--

5.6 CONCLUSION

ANN model follows the behaviour of human brain structure. In antenna designing problems, ANNs can be widely utilized for analysis and synthesis purposes on the basis of theoretical and experimental outcomes. The main aim behind this work is to analyze different kinds of antennas designed with ANN. The importance of ANN for various performance parameters is described in this paper. This technique is used for evaluating easy solutions for particular defined problems. Moreover, a comparative analysis is also done based on the antenna specifications, type of structure and application.

References

1. Kaur, M., and Sivia, J. S. (2019) Minkowski, Giuseppe Peano and Koch Curves based Design of Compact Hybrid Fractal Antenna for Biomedical Applications using ANN and PSO. *International Journal of Electronics and Communications*, 99, 14-24.
2. Kaur, M., and Sivia, J. S. (2019) ANN-based Design of Hybrid Fractal Antenna for Biomedical Applications. *International Journal of Elec-*

tronics, 106(8), 1184-1199.

3. Kaur, M., and Sivia, J. S. (2020) ANN and FA Based Design of Hybrid Fractal Antenna for ISM Band Applications. *Progress in Electromagnetics Research C*, 98, 127-140.

4. Kaur, M., and Sivia, J. S. (2019) Giuseppe Peano and Cantor set fractals based miniaturized hybrid fractal antenna for biomedical applications using artificial neural network and firefly algorithm. *International Journal of RF and Microwave Computer-Aided Engineering*, 30(1), 1-11.

5. Dhaliwal, B. S., and Pattnaik, S. S. (2017) BFO-ANN ensemble hybrid algorithm to design compact fractal antenna for rectenna system. *International Journal on Neural Computing and Applications*, 28(1), 917-928.

6. Zooghby, A. H. El., Christodoulou, C. G., and Georgiopoulos, M. (1998) Neural Network-Based Adaptive Beamforming for One- and Two-Dimensional Antenna Arrays. *IEEE Transactions on Antennas and Propagation*, 46(12), 1891-1893.

7. Guney, K., and Gultekin, S. S. (2004) Artificial Neural Networks for Resonant Frequency Calculation of Rectangular Microstrip Antennas with thin and thick Substrates. *International Journal of Infrared and Millimeter Waves*, 25, 1383-1399.

8. Gangwar, S. P., Gangwar, R. P. S., Kanaujia B. K., and Paras (2008) Resonant frequency of circular microstrip antenna using artificial neural networks. *Indian Journal of Radio & Space Physics*, 37, 204-208.

9. Thakare, V. V., and Singhal, P. K. (2009) Bandwidth Analysis by Introducing Slots in Microstrip Antenna Design Using ANN, *Progress in Electromagnetics Research M*, 9, 107-122.

10. Akdagli, A., Toktas, A., Kayabasi, A., Develi, I. (2013) An Application of Artificial Neural Network to Compute the Resonant Frequency of E–shaped Compact Microstrip Antennas. *Journal of Electrical Engineering*, 64, 317-322.

11. Can, S., Kapusuz, K. Y., and Aydin, E. (2013) Neural Network Based Resonant Frequency Solver for Rectangular-shaped shorting PIN-loaded Antennas. *Microwave and Optical Technology Letters*, 55(12), 3025-3028.

12. Dhaliwal, B. S., and Pattnaik, S. S. (2013) Artificial Neural Network Analysis of Sierpinski Gasket Fractal Antenna: A Low-Cost Alternative to Experimentation. *Advances in Artificial Neural Systems*, 2013, 1-7.

13. Sivia, J. S., Pharwaha, A. P. S., and Kamal, T. S. (2013) Analysis and Design of Circular Fractal Antenna using Artificial Neural Networks. *Progress in Electromagnetics Research B*, 56, 251-267.

14. Sivia, J. S., Pharwaha, A. P., and Kamal, T. S. (2013) Design of Sierpinski Carpet Fractal Antenna Using Artificial Neural Networks. *In-*

ternational Journal of Computer Applications, 68(8), 5-10.

15. Aneesh, M., Ansari, J. A., Singh, A., Kamakshi, and Sayeed, S. S. (2014) Analysis of Microstrip Line Feed Slot Loaded Patch Antenna Using Artificial Neural Network. *Progress in Electromagnetics Research B*, 58, 35-46.

16. Khan, T., and De, A. (2015) Modeling of Microstrip Antennas Using Neural Networks Techniques: A Review. *International Journal of RF and Microwave Computer-Aided Engineering*, 25.

17. Simsek, M. (2016) Efficient neural network modeling of reconfigurable microstrip patch antenna through knowledge-based three-step strategy," *International Journal of Numerical Modelling-Electronic Networks Devices and Fields*, 30, 185-206.

18. Gehani, A., Agnihotri, P., and Pujara, D. (2017) Analysis and Synthesis of Multiband Sierpinski Carpet Fractal Antenna Using Hybrid Neuro-Fuzzy Model. *Progress in Electromagnetics Research*, 59-65.

19. Singh, G., and Mittal R. (2017) Artificial Neural Network-based analysis of circular multiband antenna. *International Journal of Latest Technology in Engineering, Management & Applied Science*, 6(8), 75-79.

20. Khan, I., Tian, Y., Ullah, I., Kamal, M. M., Ullah, H., and Khan A. (2018) Designing of E-shaped microstrip antenna using Artificial Neural Network. *International Journal of Computing, Communications & Instrumentation Engg*, 5(1) 23-26.

21. Dhaliwal, B. S., Kaur, G., Saini, N., Pattnaik, S. S., and Josan, S. K. (2019) A Modified Two-Step ANN Ensemble Approach to Improve Generalization and its Application in Fractal Antenna Design. *Journal of Circuits, Systems and Computers*, 29(7).

22. Balanis, C. A. (2016) *Antenna Theory: Analysis and Design*, 3rd Edition, John Wiley & Sons, London.

23. Ozkaya U., and Seyfi, L. (2015) Dimension Optimization of Microstrip Patch Antenna in X/Ku Band via Artificial Neural Network. *Procedia- Social and Behavioral Sciences 195*, 2520-2526.

24. Garg, R., Bhartia, P., Bahl, I., and Ittipiboon A. (2001) *Microstrip Antenna Design Handbook*, Artech House, Boston, London.

25. Stutzman, W. L., and Thiele, G. A. (2013) *Antenna Theory and Design*, 3rd Edition, John Wiley & Sons, New York.

26. Fong, L. K. (1989) Microstrip patch antennas - basic properties and some recent advances. *Journal of Atmospheric and Terrestrial Physics*, 51(9-10), 811-818.

27. Fang, D. G. (2010) *Antenna Theory and Microstrip Antennas*, 1st Edition, CRC Press, Taylor & Francis Group, New York.

28. Shanmuganantham, T., and Raghavan S. (2009) Design of a compact broadband microstrip patch antenna with probe feeding for wireless applications. *International Journal of Electronics and*

Communications, 63(8), 653-659.

29. Bansal, A., and Gupta, R. (2018) A review on microstrip patch antenna and feeding techniques. *International Journal of Information Technology*, 12(1), 149-154.

CHAPTER 6

ENERGY AS A KEY CHALLENGE IN WIRELESS SENSOR NETWORKS

Sukhwinder Sharma[1], Puneet Mittal[2], Shreekumar T[3], Rakesh Kumar Bansal[4] and Savina Bansal[5]
[1,2,3]*Mangalore Institute of Technology and Engineering, Mangalore, India*
[4,5]*GZSCCET Maharaja Ranjit Singh Punjab Technical University, Bathinda, India*

The development and distribution of Wireless Sensor Network (WSN) has grown exponentially in recent years due to the availability of affordable, low-power devices such as processors, sensors, and radios, which are often integrated into a single chip. As a result of advances in VLSI, computer, communications, and micro-electromechanical technologies, researchers' interest has grown exponentially in addressing and improving communication capacity and processing data-restricted resources. WSNs are emerging as a way of collecting information to build a knowledge-building and communication system, which will significantly improve the efficiency and reliability of infrastructure. Trouble-free deployment and greater flexibility of devices are additional benefits of WSN over traditional solutions. With the swift progress in the development of sensors, WSNs will shortly develop as the main technology behind the Internet of Things (IoT). WSN technology provides an effective solution to allow ubiquitous access to online monitoring throughout the industrial process, therefore, industrial automation is turning out to be one of the leading areas of WSN applications. In addition, technology plays a key role in monitoring safety beyond the transmission lines and transformation equipment [1]. The variety of WSN applications in the real world is limitless and is slowly increasing from the previous defense to the industrial and commercial sectors (Figure 6.1).

Wireless Sensor Network Debut

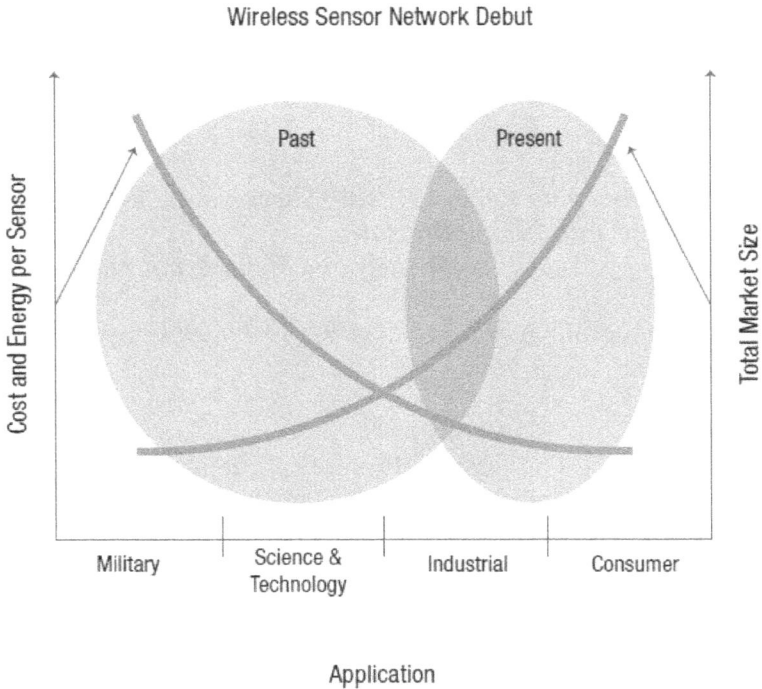

Figure 6.1 Market size of WSN applications [2].

According to [3], the WSN market worth USD 29.06 billion in 2016 is projected to range around USD 93.86 billion by 2023, with a combined annual growth rate of 18.55% during the period between 2017 and 2023. Looking at the strengths of WSNs, it has been a hotbed of research over the past decade and the dedicated efforts of researchers have led to significant contributions and advances in nearly all aspects, including software, hardware, program design, system architecture, tool support, standards, and usage.

Due to increased application requirements, too much is being demanded of resource-limited sensor nodes. To fulfill future demands, various design issues need attention and healthy efforts to meet related challenges. Power efficiency is the most critical issue for WSNs, where each sensor node depends on its battery capacity limited to data sensitivity, performance and connection (transmission and reception). Due to the transmission power limit, available bandwidth and wireless radio frequency range are often limited. Since the sensor nodes are usually very small, powered by a non-rechargeable battery, power control becomes a major and most challenging problem in WSN design [4,5]. In addition, each sensor node has a different level of power consumption due to the uneven sensing of the event and the distance from the sink. This leads to power divisions between sensory areas in the network that reduce network life. Batteries need to be changed or recharged regularly, which might be difficult due to demographic conditions. Therefore, limited power resources need to be utilized very judiciously by incorporating energy efficiency aspects in each and every hardware and software design component of WSNs.

6.1 INTRODUCTION

WSN (Figure 6.2) contains a huge number of self-configuring nodes, small and inexpensive devices with a radio transceiver, spanned over a wide sensing field, so that, monitoring without human communication is possible in the environment where infrastructure such as power, internet connection is not available. Sensor node having one or more sensors collects the sensed data, analyzes it and transmits it to the sink or base station directly or via other nodes. Advances made in the field of Microelectromechanical Systems (MEMS) and Nanoelectromechanical Systems (NEMS) over the recent years have facilitated the development and availability of a wide variety of sensors in the market [6], which are accomplished to sense a wide range of parameters like humidity, pressure, velocity, stress, strain, sound, light, heat, temperature, pH value, chemical composition and many more from the field. After that, sensed data processing is done at various levels and is then communicated to sink. Sink node has higher processing and transmission capabilities to transfer the processed information to the end-user at some remote location via internet or any other networking infrastructure. The information is finally used to monitor and control the physical phenomenon that is under the consideration of WSN.

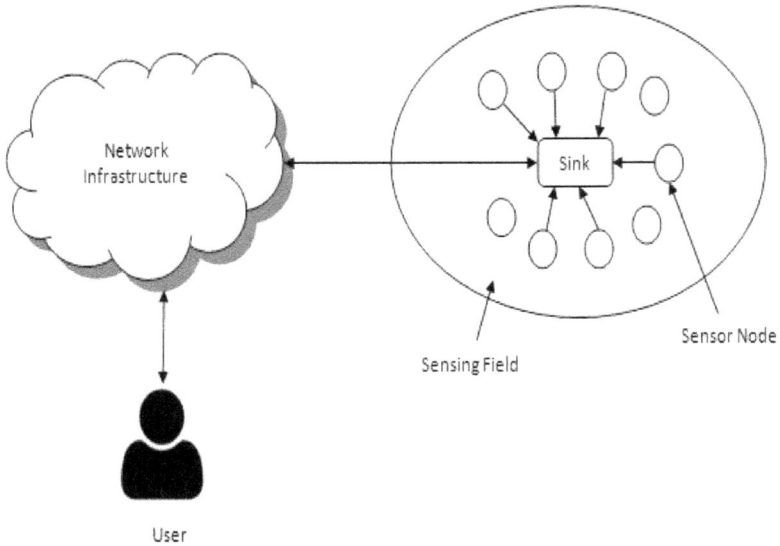

Figure 6.2 Wireless sensor network.

Due to the miniaturization issues of sensor areas, these nodes have a limited capacity battery thus interfering with applications that require extended battery life. The sensor node can be supported by energy harvesting technology [7], as it requires very little power for its operation, thanks to MEMS/NEMS technology. However, the energy generated by the environment depends on natural conditions and is rarely sustained. For a few services, the battery can be replaced or renewed. But for most applications, where the sensing field is located in some hard to reach place like oceans or mountains, recharging or replacing the battery is not possible. However, wireless technology and control technology is used to fill the gap between humans and electronics. Quick deployment, self-organizing, high reliability, affordability, flexibility and tolerance features of WSN makes them a promising way to be used in various sensing applications. With the help of these networks, it is very helpful to gather information in places that are not easily accessible and difficult to access. WSNs hold the power to provide a low-cost solution to the challenges being faced in the war field, military surveillance, climate/weather conditions and medical diagnoses to name a few. In crux, WSN should automatically do various tasks of sensing and monitoring in difficult environments and situations like forest fires, avalanches, Tsunamis etc., so that, disastrous situations can be prevented.

6.1.1 Sensor Node

A sensor node is a small device having essential components like sensors, processors, radio transceiver, converters, I/O ports, memory, battery and/or additional components like location finding system, and mobilization (Figure 6.3). Being a self-sufficient system, its operation can be divided into three subsystems: i) a sensing subsystem that acquires data from the physical environment, ii) a processing and storage subsystem which processes and stores data, and iii) a communication subsystem that transmits and receives data and provide external storage of data. Additionally, the power source is also there on the sensor node for energy supply, so that the device can perform tasks. Generally, this power source contains a battery with limited energy. AA Alkaline battery is quite popular among sensor node manufacturers due to its characteristics suitable for WSNs [8].

Sensing Subsystem: It is used for collecting data samples from the surrounding environment of the physical entity, that is to be sensed and monitored, and converts into a suitable electrical signal for further processing. It may have more than one sensor to sense different physical parameters [9]. The responsibility of the sensing subsystem is to collect samples at a pre-defined interval of time and to pass it on to the processing and storage subsystem via analog and digital I/O ports. Some sensors generate digital signals, which can be directly processed, while analog signals need to be converted into digital signals with the help of an A/D converter and then processed.

Processing and Storage Subsystem: This section is basically used for storing and further processing digital samples received from the sensing subsystem. Due to less power consumption and more flexibility, micro-controllers are the most suitable choice for sensor nodes [10]. Micro-controllers perform the main task pertaining to the processing of data and controlling the functions of other components. It is used to process and store digital samples, and forward these to the communication subsystem as and when required. Micro-controller consists of essential components like processor, internal memory, and analog/digital to digital/analog converters, and additional components like location finding system, mobilizer, and actuators depending on the type of application for which a sensor node is used for. Depending upon the application requirements, the position of the sensor node may be detected using the location finding Global Positioning System (GPS). A mobilizer can be used to relocate the sensor node spatially.

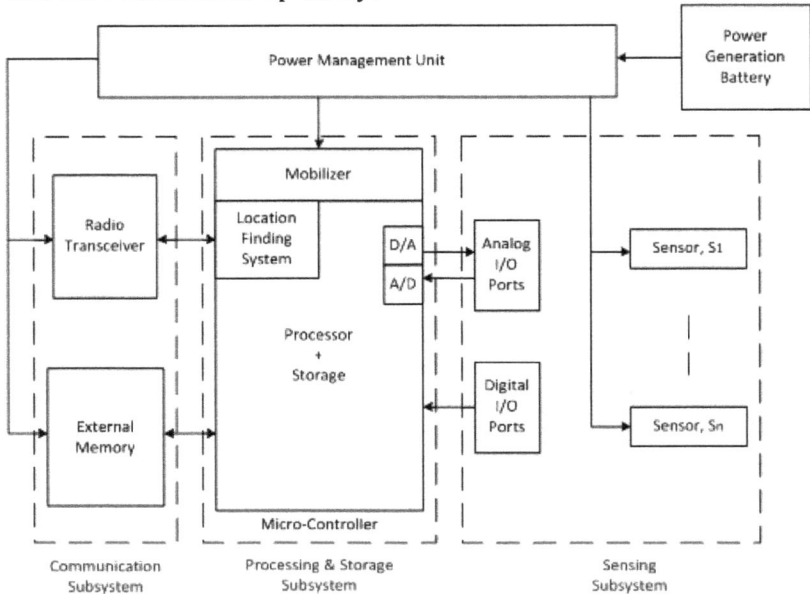

Figure 6.3 Components of a typical sensor node.

Communication Subsystem: It is used to communicate with neighboring nodes and sinks. It comprises a short-range radio transceiver and a limited capacity external memory for data persistence. Sensor nodes have to store and maintain databases for real-time traffic comprising of their samples and data received from other nodes depending upon applications and network topology. For this purpose, it requires external memory and efficient memory management. Generally, EEPROM or flash memory is used to store data in the communication subsystem. Depending upon the range of transceiver radio, the sensor node may send data directly or through other nodes to the sink.

Power Management Unit: To energize various subsystems of a sensor node, this unit supplies energy to all components as and when required. For better and efficient power management, it might be required to keep a sensor node in an idle, active or sleep state. Further, owing to miniaturization constraints, size should not exceed a certain limit. For some nodes, the size may be less than a cubic centimeter. Accordingly, due to these deployment and size constraints, a limited capacity battery is generally installed, which may function in high volumetric densities, be dispensable and autonomous, operate without human interventions, and can adapt in any environment [11]. Judicious consumption of available energy is one of the key design issues for WSNs. Keeping in view the wide application range of WSNs, and depending upon the application characteristics and field size, different kinds of sensor nodes are deployed to sense, process, store and communicate data. Commercial products available in the market like COOKIES, BTnode, EPIC mote, Eyes, FlatMesh FM1, IMote, IMote 2.0, T-Mote Sky, Waspmote, INDriya, Iris Mote, Mica2, MicaZ, Mulle, PowWow, Preon32, Shimmer, and TelosB are some of the popular sensor nodes [12,13].

6.1.2 Sink Node

Sink node is a resource-rich device that accumulates the data from sensor nodes and sends the processed information to the end-user through the underlying communication network. It forwards user queries to the appropriate node in the network and sends aggregated and summarized sensor data to users. Generally, in literature, it is assumed that the sink node has high-speed processing, long-distance communication and large storage capabilities with a sufficient amount of energy, which cannot be depleted during the network operation. Sink nodes are either directly connected to power supply or to some ambient energy harvesting technique or other mechanisms that may include micro oscillators, electromagnetic wave reception devices, miniature piezoelectric crystals and thermoelectric power generation elements [7,8] to recharge their batteries. Sink takes queries from end-user and responds back giving desired information from the sensing environment. Most of the WSN systems use a *n-to*-1 communication paradigm in which sensor nodes forward their data towards a common static sink deployed at the centre, corner or some pre-specified location on the sensing field. On the other hand, the mobile sink can be deployed to evenly distribute the network load for increased network lifetime and to reduce the bottleneck problem of static sinks [14]. Multiple sinks can also be deployed to spread the load over the network. Meshlium [15] and CrossBoweKo [16] are commercially available devices that can act as a sink for WSNs. Finally, the underlying wired or wireless network provides communication between the sink and the end-user. The end-user can monitor the sensing field through data received from the sink at predefined intervals or may send queries to fetch the desired information from the sink.

6.2 DATA COLLECTION IN WSNs

WSN aims to provide observations of a sensing field where the value of an observation is important for the end-user rather than the process through which data is being collected, thereby makes the characteristics of WSNs data-centric [17]. Due to the data-centric nature of WSNs, data collection stages a significant role to provide fast and effective communication of useful information to the end-user with efficient sensing and processing of data samples at the sensor nodes.

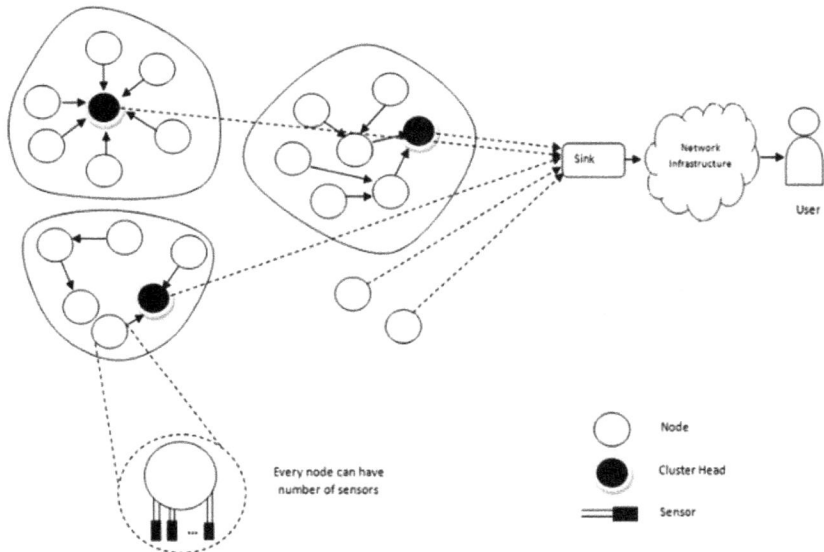

Figure 6.4 Data collection process in WSN.

The data collection process (Figure 6.4) involves three elementary operations: sensing, processing and communication of sensed data to the sink. The sink further processes this data and forwards processed data to the end-user through network infrastructure like the internet for decision making. Depending upon the network topology, data collection techniques can be divided into four categories: (i) direct communication based, (ii) chain based, (iii) clustering-based, and (iv) hybrid.

6.2.1 Direct Communication Based Techniques

In direct communication techniques (Figure 6.5), each sensor node sends data straight to the sink. If the sink is distant from the nodes, direct communication will involve a huge amount of energy depletion from each node. This will rapidly trench the batteries of nodes and diminish the system lifetime. This scenario is suitable for time-critical and security-based applications, where indirect communications may lead to unnecessary delay and security threats to the data being transmitted. Flat networks are the example of direct communication-based networks. Due to direct communication, there is unnecessary communication in flat networks, and they put additional load on sink node which results in earlier exhaustion of the battery. Sink being the only receiver makes these networks fast and secure, therefore these techniques can be used for high speed and security-aware applications within a small geographical area.

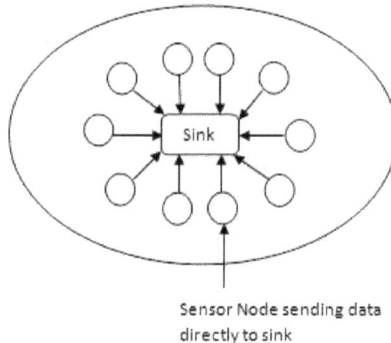

Sensor Node sending data
directly to sink

Figure 6.5 Direct communication techniques.

6.2.2 Chain Based Techniques

If the sink is not within the transmission range of nodes, direct communication is not possible. One alternative for this situation is the chain based technique (Figure 6.6).

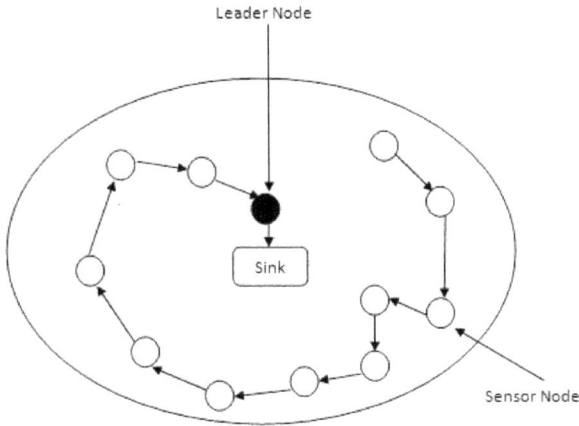

Figure 6.6 Chain based techniques.

In chain based networks, a chain of nodes is formed for sending the data to the sink. Each sensor node sends data to its neighboring node and neighboring node upon receiving the data, adds its own data to it and sends it to another neighboring node (excluding the node which sent data to it), forming a chain. This is continued until data reaches its chain leader. Then leader adds their data into the packet and sends it to the sink.

6.2.3 Clustering Based Techniques

For huge networks, it is unproductive to route data directly to the sink. Another alternative to direct communication is the clustering based techniques. In clustering based techniques, the network is divided into clusters, where each cluster has a Cluster-head (CH) and associated sensor nodes. The sensor nodes sense the environment and communicate the sensed data to the sink node through one or more CHs in a single-hop or multi-hop manner as shown in Figure 6.7.

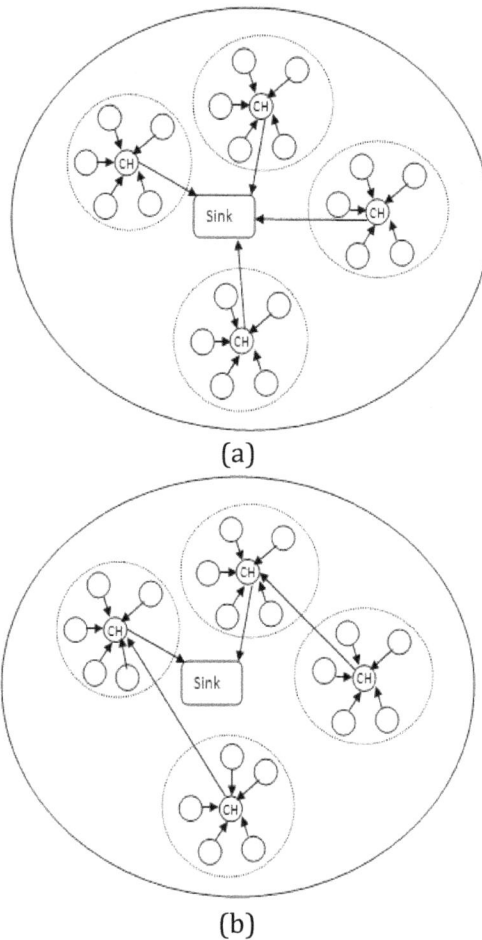

(a)

(b)

Figure 6.7 Clustering based communication: (a) Single-hop, (b) Multi-hop.

Clustering based techniques are efficient in two ways: firstly, it reduces the transmission distance of nodes by limiting them to transmit data to the CH rather than sink. Secondly, data received from associated nodes can be processed or aggregated at the CH which reduces the amount of data to be communicated to the sink. Therefore, it provides long-distance communications to the resource-poor nodes as well as balances the load of resource-poor networks. Due to their promising advantages, a large number of clustering based techniques have been proposed by eminent researchers in the field. LEACH (Low Energy Adaptive Clustering Hierarchy) protocol [18] and its variants are the popular choices among homogeneous WSNs while different clustering based techniques have been designed for heterogeneous WSNs having nodes with different capabilities. The major design issues for these techniques are to decide an optimal number of clusters, efficient CH selection process, and aggregation of sensed data at various levels.

In single-hop communication, there is only one intermediate CH while multi-hop communication having multiple CHs may be adopted for data transmission. While single-hop communication may be used for applications with a small geographical area having sparsely distributed nodes, multi-hop communication may be used for applications with a large geographical area having densely distributed nodes.

6.2.4 Hybrid Techniques

Depending upon the underlying network topology two or more techniques can be combined to get the best suitable alternative for the application-specific scenario. While nodes at the network boundaries or highest distant nodes from the sink may opt for chain based or cluster-based data collection techniques, nodes near the sink may opt for direct communication to efficiently utilize available resources as well reduce the burden of intermediate nodes or CHs.

6.3 RESEARCH ISSUES AND CHALLENGES IN WSNs

The ultimate goal of a sensor network is to transmit reliable information to the end-user through the sink at desired quality level with efficient utilization of available resources. Various key issues pertaining to the design and performance of WSN are listed below:

Energy: Energy efficiency is the gravest design issue in WSNs, where each sensor node relies on its limited battery power for data sensing, processing, and communication (transmission and reception). Available bandwidth and a radio range of wireless channels are often limited. Due to the small size of sensor nodes and irreplaceable battery, energy control is a quite important and challenging design issue [19]. A lot of energy disparity is observed in the network due to the varying energy consumption rate of sensors, their distance from the sink node and unequal event sensing periods. This results in a shortening of the lifetime of the network. Batteries need to be changed or recharged regularly, which might be difficult due to demographic conditions. Therefore, limited power resources need to be utilized very judiciously by incorporating energy efficiency aspects in each and every hardware and software design component of WSNs.

Deployment: Deployment relates to the physical implementation of the network in the real scenario. It is a very arduous and cumbrous activity. It largely relies on the demographic location of the application. At remote locations, sensors are thrown from arial vehicle or are placed based on some predefined topology. Coverage and connectivity are the two major challenges for node deployment in WSNs [20].

Self Management: Self management is a complex issue due to the large scale of networks with resource restrictions and dynamic conditions [21]. Depending upon the applications, sensor nodes can be dropped over remote locations like oceans, dense forests, volcanoes. Once sensor nodes are deployed they are expected to function freely, without any human interference, for years. The network is expected to function, manage, maintain and repair by itself after deployment.

Hardware and Software Issues: Due to size, energy and deployment constraints, sensor nodes have limited hardware and software capabilities. Integration of hardware components like microcontroller, battery, memory chips and sensors for a small-sized node requires the production of out of the ordinary hardware. Special attention is required for designing simple operating systems for WSN to protect the inner details of sensor nodes from the application by providing an interface to the external world [22]. Operating System has to allocate the limited resources in a correct and controlled manner. Contiki, TinyOS, Nano-RK, MANTIS and LiteOS are some of the popular operating systems meant for WSNs. Energy-aware policies are needed in every aspect of memory management, task scheduling, resource sharing, reprogramming, and software upgradations [23].

Synchronization and Calibration: Network synchronization provides a standard time zone for local nodes in a network. The global clock in the sensor system will help to process and analyze data efficiently and predict future system performance. Other systems that require global clock synchronization are environmental monitoring, navigation, traffic tracking etc. The power consumption of some synchronization schemes is due to power-hungry devices such as Global Positioning System (GPS) receivers or Network Time Protocol (NTP) receivers. Sensors need to be synced because it can lead to incorrect data measurement. The design of efficient syncing systems with high accuracy and low resource utilization is a major challenge for syncing in WSNs [24, 25].

Quality of Service: Service quality (QoS) is a service level provided by sensory networks to its users. WSNs are used for a variety of occasions and critical systems, so it is imperative that the network provides good QoS. However, it is hard because the network topology is continually changing and the available state traffic information is unclear. Sensitive networks need to be provided with the required amount of bandwidth in order to achieve the required minimum QoS. Traffic cannot be measured in the sensor network because data is aggregated from multiple nodes to node nodes. QoS methods require unlimited QoS traffic to be built. Most of the time travel on sensory networks requires the provision of energy efficiency to meet delivery needs. Although multi-hops reduce the amount of power used for data collection, the accompanying hop can reduce packet delivery. The QoS designed for WSN should be able to support disability [26].

Security: WSNs are used not only for military applications but also for recruitment, structure surveillance, intruder alarms and critical systems like hospitals and airports, therefore, security is a major problem. Confidentiality is required for sensory networks to protect information that travels between network sensors or between sensors and the sink; otherwise, it may lead to a communication hearing. For sensory networks, it is important that each node of the grief and sink has the power to ensure that the received data is actually sent by a trusted sender and not the enemy who has deceived the official sites into receiving false data. False data can change the way the network can be predicted. Data storage should be maintained. Data should not change and accurate data should reach the end-user. Various types of threats to sensory networks manipulate and distort route information, untrained data collection, node destruction, sybil attacks, Denial of Service attacks, sinkhole attacks and jamming [27].

Decentralized Management: Large and powerful barriers make it difficult to rely on intermediate algorithms as it is done in one business to create network management solutions such as topology and/or routing. Instead, node sensors must work with their neighbors to make localized decisions, that is, without global knowledge. As a result, the results of distributed (or distributed) algorithms will not be correct but may be more robust than intermediate solutions due to shorter headaches.

MAC and Routing Protocols: The development of Medium Access Control (MAC) protocols has a direct impact on energy consumption, as some of the main causes of energy waste are found in the MAC layer: collisions, control packet overhead and inactive listening. The advanced energy-saving process for error management is not easy to use due to its high computing power requirements and the fact that long packets often do not work. The range of low sensors leads to dense networks so it is necessary to find an effective medium access process depending on the power limits. Navigating WSN is a major challenge due to the environmental factors that distinguish these networks from other wireless networks such as mobile ad-hoc networks or mobile networks. First, due to a large number of sensor nodes, it is not possible to create a global communication system for the deployment of a large number of sensor nodes as more ID storage is higher. Therefore, traditional IP-based protocols may not be applicable to WSNs. In addition, temporary sensor nodes need to be configured as the ID distribution of these entities requires the system to make connections and deal with the distribution of nodes which has led to the neglect of sensory networks [28].

Robustness: The sensor network must remain active even if any node fails while the network is active. The network should be able to adapt by changing its connection in case of errors. If so, an efficient algorithm is used to completely change the network configuration. To support the desired lifetime needs, each node must be built to be as strong as possible. For standard deployments, hundreds of nodes will need to be compatible for years. To achieve this, a system must be built to be able to tolerate and adapt to the failures of each node. In addition, each node should be designed to be as powerful as possible. System modularity is a powerful tool that can be used to create a robust system. By dividing a program's performance into individual pieces, each task can be thoroughly evaluated on its own before merging them into a complete program. To make this easier, parts of the system should be as independent as possible and have smaller connections, preventing unexpected connections. In addition to increasing system stability at node failures, WSN should also be effective in external interference. Since these networks often interact with other wireless systems, they need the ability to adapt to their behavior appropriately. The severity of wireless connectivity for external interference can be greatly increased through multichannel radios and distribution. It is common for buildings to have existing wireless devices running on long or multiple frequencies. The ability to avoid congested frequencies is important to ensure successful deployment.

Real-Time and Multimedia Communication: Most real-time WSNs must achieve real-time performance in the longest lifetime. While energy harvesting has shown promise as a technology that empowers long-term WSNs, it also presents new challenges in real-time processor scheduling due to the flexibility of energy resources and limited energy storage capacity. Multimedia data is collected and transmitted by the sensor network. In addition to standard data delivery methods for scalar sensor networks, multimedia data includes summary and streaming of multimedia content. The processing and delivery of multimedia content are independent and their communication has a significant impact on accessible QoS. They want higher transmission bandwidth.

6.4 ENERGY CONSUMPTION AS A KEY CHALLENGE IN WSNs

Energy is the most crucial design issue in WSNs needs special attention. This issue has been sincerely taken up by various researchers. While some researchers have investigated the major sources of energy consumption, others have proposed data collection techniques having efficient energy consumption through different algorithms. Depending upon the sensing, processing and communication subsystems, a sensor node having one or more sensors allocates its energy for network operation at four different levels: sensor, node, cluster and network levels. At the sensor level, energy is used up by sensors while sensing data from the environment; at the node level, energy is consumed during collection, compression or manipulation of data sensed by sensors; at the cluster level, energy is consumed during data aggregation between multiple nodes of a cluster; and at the network level, energy is consumed between multiple CHs, relays and sink(s). Along with this, energy is consumed due to the mobility of sensor nodes, mobile relays and sinks, in the case of mobile WSNs.

6.4.1 Sources of Energy Consumption

As discussed earlier, the energy is consumed during sensing, processing and communication operations. Depending upon the nature of applications and type of sensors, the consumed energy per sample varies from 0.0048 mJ to 225,000 mJ [29] for sensing operations. The energy consumption of various sensors is shown in Table 6.1.

Table 6.1 Energy consumption of sensors [29]

Sensor Type	Representative Sensor	Sensing Energy Consumption (mJ)
Acceleration	MMA7260Q	0.0048
Pressure	2200/2600 Series	0.0225
Light	ISL29002	0.123
Proximity	CP18	48
Humidity	SHT1X	72
Temperature	SHT1X	270
Level	LUC-M10	1660
Gas (VOC)	MiCS-5521	4800
Flow Control	FCS-GL1/2A4-AP8X-H1141	17500
Gas (CO2)	GE/Telaire 6004	225000

As per the first-order energy consumption model [30], processing operations consume 5nJ/bit/signal energy for data aggregation and compression operations. Communication operations involve transmission and reception of data, where 50 nJ/bit energy is consumed to run the transmitter/receiver electronics; while 10 pJ/bit/m^2(free-space) and 0.0013 pJ/bit/m^4(multi-path) energy is consumed to run the transmitter amplifier. As per existing literature, communication functions require a major portion of energy for transmission and reception of data to and from other nodes/sink. It also includes energy consumed by the transmitter amplifier to sleep, listen and switch between sleep and listen to modes. Energy consumed by different sensor nodes for different communication functions is shown in Table 6.2.

Table 6.2 Energy consumption for communication [29]

Components of E_{comm}	CC1000	TDA5250	CC2420	AT86RF230
E_{rx}(millijoules)	19.62	97.3	6.38	4.83
E_{tx}(millijoules)	52.97	18.83	5.97	5.13
E_{listen}(millijoules)	13.83	85.7	30.13	22.12
E_{sw}(millijoules)	194.4	669.6	136.54	172.73
E_{slp}(millijoules)	1.078	0.0054	1.077	6.47
E_{comm}**(millijoules)**	**281.87**	**871.45**	**180.10**	**211.29**

Table 6.3 Energy consumption (Communication v/s Sensing and Processing) [29]

Sensors	Mica2		TelosB		Imote2	
	E_{sm}	E_{comp}	E_{sm}	E_{comp}	E_{sm}	E_{comp}
2200/2600 Series	0.000079	0.096	0.00013	0.044	0.00013	4.01
CP18	0.17	0.105	0.267	0.047	0.267	4.12
SHTIX(H)	0.255	0.77	0.4	0.043	0.4	12.8
SHTIX(H)	0.957	2.65	1.5	0.94	1.5	37
MiCS-5521	17.242	5.2	26.98	1.84	26.98	69.9
GE/Telaire 6004	798.2	25.64	1249.25	9.03	1249.25	333.8
MMA-7260Q	0.000017	0.096	0.0000268	0.044	0.0000268	4.01
ISL29002	0.00044	0.106	0.00068	0.047	0.00068	4.13
LUC-M10	5.89	0.266	9.22	0.104	9.22	6.2
FCS-GL1/2A4-AP8X-H1141	62.1	1.28	97.2	0.46	97.2	19.4

To understand the relationship between energy consumption for communication operations as compared to sensing and processing operations, Table 6.3 shows the energy consumed for communication operations by some standard sensor nodes having the above mentioned sensors. It shows how many times the energy is consumed by the sensor node for sensing and processing operations as compared to communication operations. It can be observed that sensing cost primarily depends upon the type of sensors installed on the sensor node. For most general-purpose sensors, the sensing cost for each data sample is much lower than the communication cost. Further, processing costs are higher than sensing costs but much lower than communication costs involved. As per the literature surveyed [31], it is gathered that almost half of the total energy gets consumed during data communication (Figure 6.8). Hence, efficient techniques for managing data communication need more emphasis on the effective utilization of the limited low-capacity battery.

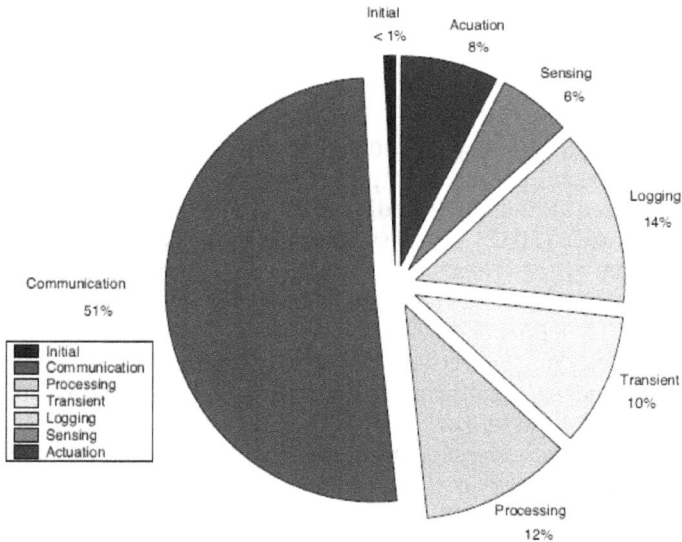

Figure 6.8 Sources of energy consumption [31].

6.5 SUMMARY AND CONCLUSIONS

This chapter gives an overview of the key components, processes and sources of energy consumption in wireless sensor networks. Energy, being the most prominent challenge in WSNs, is discussed in detail giving energy requirements in different sensor node subsystems and data collection process phases. Initially, the development and applications of WSN are discussed. Then, the WSN components-sensor node and sink node are given, where different sensor node components and their energy consumption process is elaborated. The process of data collection involving various subsystems is explained describing the energy consumption in data sensing, aggregation and communication subsystems. Afterwards, the issues and challenges faced by WSNs are discussed. Finally, Energy being the key challenge in WSNs is discussed to understand the sources of energy consumption, the energy required for individual sensor types, and the relationship between different key operations. This work is useful to the researchers who are working on different energy-efficient protocols and procedures to cut down energy consumption to the possible extent. It is easier to improve energy efficiency if the sources, processes and system components involved in energy consumption are well known.

References

1. Yinbiao, S., and Lee, K. (2014) Internet of Things: Wireless Sensor Networks. Online: http://www.iec.ch/whitepaper/pdf/iecWP-internetofthings-LR-en.pdf (accessed 22nd January 2021).
2. Cooley, D. (2012) Wireless Sensor Networks Evolve to Meet Mainstream Needs. *RTC Magazine*, Online: https://issuu.com/rtcgroup/docs/rtc1212 (accessed 22nd January 2021), pp. 26-38.
3. Wireless Sensor Network Market by Building Automation, Wearable Devices, Healthcare, Automotive, Transportation-2023, MarketsandMarkets, (n.d.). Online: https://www.marketsandmarkets.com/Market-Reports/wireless-sensor-networks-market-445.html (accessed 22nd January 2021).
4. Sharma, S., Bansal, R. K., and Bansal, S. (2013) Issues and Challenges in Wireless Sensor Networks. *International Conference on Machine Intelligence and Research Advancement*, Katra, India, pp. 58-62.
5. Pantazis, N. A., and Vergados, D. V. (2007) A Survey on Power Control Issues in Wireless Sensor Networks. *IEEE Communications Surveys & Tutorials*, **9**(4), 86–107.
6. Lee, C. (2013) MEMS/NEMS Based Enabling Technologies for Self-Sustained Wireless Sensor Nodes. *IEEE MTT-S International Microwave Workshop Series on RF and Wireless Technologies for Biomedical and Healthcare Applications (IMWS-BIO)*, Singapore, pp. 1–3.
7. Zhou, G., Huang, L., Li, W., and Zhu, Z. (2014) Harvesting Ambient Environmental Energy for Wireless Sensor Networks: A Survey.

Journal of Sensors, **2014**, 1–20.
8. Knight, C., Davidson, J., and Behrens, S. (2008) Energy Options for Wireless Sensor Nodes. *Sensors*, **8**(12), 8037–8066.
9. Waspmote - Open Source Sensor Node for the Internet of Things, (n.d.), Online: http://www.libelium.com/products/waspmote (accessed 22nd January 2021).
10. Aponte-Luis, J., Gómez-Galán, J. A., Gómez-Bravo, F., Sánchez-Raya, M., Alcina-Espigado, J., and Teixido-Rovira, P. M. (2018) An Efficient Wireless Sensor Network for Industrial Monitoring and Control. *Sensors*, **18**(1), 1–15.
11. Asim, M. (2010) Self-Organization and Management of Wireless Sensor Networks. *Ph.D. thesis*, Liverpool John Moores University, U.K.
12. Sharma, S., Bansal, R. K., and Bansal, S. (2016) Chapter 11: Energy-efficient data collection techniques in wireless sensor networks. In: *Emerging Communication Technologies Based on Wireless Sensor Networks: Current Research and Future Applications*, CRC Press, Taylor & Francis, USA, pp. 275-296.
13. Ciabattoni, L., Freddi, A., Longhi, S., Monteriù, A., Pepa, L., and Prist, M. (2016) An Open and Modular Hardware Node for Wireless Sensor and Body Area Networks. *Journal of Sensors*, **2016**, 1–16.
14. Thanigaivelu, K., and Murugan, K. (2009) Impact of Sink Mobility on Network Performance in Wireless Sensor Networks. *IEEE 1st International Conference on Networks and Communications*, Chennai, India, pp. 7–11.
15. Meshlium Xtreme - The Internet of Things IoT, (n.d.), Online: http://www.libelium.com/products/meshlium/ (accessed 25th January 2021).
16. Crossbow eKo, (n.d.) Online: http://afriweather.co.za/crossbow_eko.htm (accessed 25th January 2021).
17. Jallad, A., and Vladimirova, T. (2009) Data-Centricity in Wireless Sensor Networks. *Guide to Wireless Sensor Networks*, Springer, London, pp. 183–204.
18. Heinzelman, W. R., Chandrakasan, A., and Balakrishnan, H. (2000) Energy-Efficient Communication Protocol for Wireless Microsensor Networks. *33rd IEEE Annual Hawaii International Conference on System Sciences*, USA, pp. 1–10.
19. Pantazis, N. A., and Vergados, D. V. (2007) A Survey on Power Control Issues in Wireless Sensor Networks. *IEEE Communications Surveys & Tutorials*, **9**(4), 86–107.
20. Liu, Y. (2012) Wireless Sensor Network Applications in Smart Grid: Recent Trends and Challenges. *International Journal of Distributed Sensor Networks*, **8**(9), 1–8.
21. Ruiz, L. B., Braga, T. R. M., Silva, F. A., Assuncao, H. P., Nogueira, J. M.

S., and Loureiro, A. A. F. (2005) On the Design of a Self-Managed Wireless Sensor Network. *IEEE Communications Magazine*, **43**(8), 95–102.

22. Gupta, K., and Sikka, V. (2015) Design Issues and Challenges in Wireless Sensor Networks. *International Journal of Computer Applications*, **112**(4), 26–32.

23. Farooq, M. O., and Kunz, T. (2011) Operating Systems for Wireless Sensor Networks: A Survey. *Sensors*, **11**(6), 5900–5930.

24. Rhee, I., Lee, J., Kim, J., Serpedin, E., and Wu, Y. (2009) Clock Synchronization in Wireless Sensor Networks: An Overview. *Sensors*, **9**(1), 56–85.

25. Tan, R., Xing, G., Yuan, Z., Liu, X., and Yao, J. (2010) System-Level Calibration for Data Fusion in Wireless Sensor Networks. *31st IEEE Real-Time Systems Symposium (RTSS)*, CA, USA, pp. 215–224.

26. Xia, F. (2008) QoS Challenges and Opportunities in Wireless Sensor/Actuator Networks. *Sensors*, **8**(2), 1099–1110.

27. Wang, Y., Attebury, G., and Ramamurthy, B. (2006) A Survey of Security Issues in Wireless Sensor Networks. *IEEE Communications Surveys & Tutorials*, **8**(2), 2–23.

28. Al-Karaki, J. N., and Kamal, A. E. (2004) Routing Techniques in Wireless Sensor Networks: A Survey. *IEEE Wireless Communications*, **11**(6), 6–28.

29. Razzaque, M., and Dobson, S. (2014) Energy-Efficient Sensing in Wireless Sensor Networks Using Compressed Sensing. *Sensors*, **14**(2), 2822–2859.

30. Callaway, E. H. (2003) *Wireless Sensor Networks: Architectures and Protocols*, Auerbach Publications, London, England.

31. Halgamuge, M. N., Zukerman, M., Ramamohanarao, K., and Vu, H. L. (2009) An Estimation of Sensor Energy Consumption. *Progress In Electromagnetics Research B*, **12**, 259–295.

CHAPTER 7

RECOMMENDER SYSTEM FOR IDENTIFYING POTENTIAL CUSTOMERS USING MACHINE LEARNING

Navdeep Singh and Kulwinder Kaur
CSE, Punjabi University, Patiala, India

7.1 INTRODUCTION

Today, giving a computer some operation requires a particular set of directions or a coded algorithmic program which describes the procedures to be executed. The systems currently lack the ability to learn from previous experiences and so it is difficult to improve on the basis of prior failures. As a result, teaching a computer to do a job with a controlled programme required the creation of a precise and efficient algorithm, which is a time-consuming endeavour whereas, in Machine Learning, a large quantity of data is analysed and is used for doing future predictions and guide companies to shape their strategy by using predictions [1]. The model projections help companies shape their strategy by using real-life data. Decisions are made by algorithms based on past data and statistical representations [2]. Machine learning, a subset of AI has emerged as a section of the continuing quest to develop intelligent systems that may learn from their experience using the most representative data set [3]. A machine learning solution consists of three main components: the data, the obtained features from the data and the model. The experience gained over the previous decade has proven that in real life, the size of the dataset is the most important element [4]. It has also evolved from the need to teach computers how to automatically learn a solution for a problem, and the system gets smarter over time. It utilizes a statistical algorithm that trains and develops on its own without the assistance of humans [5]. Machine learning generates an algorithm from subsets of data that can utilise combinations of features and weights different from those obtained from basic principles. In brief, it is the study of strategies for developing and improving systems via the analysis of instances of their intended behaviour. Machine learning is useful in domains where humans are unable to provide precise results for the targeted program's behaviour [6]. Arthur Samuel introduced the concept of machine learning in 1959 and described it as an area of research that allows computers to learn without being explicitly taught. According to Tom Mitchell, chairman of machine learning at Carnegie Mellon University, "a computer programme is said to learn from experience E, task T, and evaluation measurement P, if its achievement on task T, as estimated by P, increases with experience E" [7]. In 2016, Haffner stated it aptly "machine learning is made up of algorithms that teach computers to execute activities that humans do easily and readily on a regular basis" [8]. A computer programme is assigned to execute specific tasks in machine learning, and the system is considered to have learnt from its experience if its observable performance on these tasks increases as it gets more experience by executing these tasks. Based on learned facts, the machine then makes judgments and predictions [9]. There are mainly four learning methods that are used to perform tasks which are supervised learning, semi-supervised, unsupervised, and reinforcement learning [6]. Extensive work has already been done in different fields utilising machine learning, some areas are discussed below:

• When a photo is uploaded on social media or something is purchased from online selling websites, or a series is browsed on Netflix, almost all of these sites collect data and fingerprints about us. All produced relevant factors are gathered and analysed, and machine learning assists in making sensible suggestions. YouTube/Netflix may upgrade the playlist according to the sort of videos that are watched, and Facebook can propose posts by observing what type of goods we buy regularly; online selling websites may advise our subsequent choice based products by considering the wallet size [2].

• By analysing the results of a person's medical survey, computer software can help us to forecast whether a person has cancer or not. As more expertise is acquired evaluating medical survey reports from a larger patient group, the performance level of the model increases. The performance is assessed by the number of right forecasts and cancer cases detected that have been confirmed by a skilled oncologist [9].

• Facial recognition is another example. The system can recognise relatives and friends by detecting the faces or pictures, regardless of variations in posture, face tone, hair shade etc. A facial picture, as we are all aware, is made up of symmetrical arrangements. The eyes, nose, and mouth are all situated at different locations on the face. Every individual's face is formed up of a shape made up of a certain composition, and a learning software records that person's distinctive pattern and then recognises it by verifying the pattern on a specific image by trying to analyse sample face photos of that person. It is an appropriate illustration of pattern recognition [1].

• In addition, attendance recognition system, email as spam, unique recognition of the characteristics of a certain fruit, banks with facial ageing and facial recognition, google home for voice help, translator search engines, internet search engine behaviour, fraud credit card alerts, tax filing anomaly detection, online product buying advice, voice recognition, facial recognition, Amazon Alexa, driver assistance features, YouTube, Netflix for recommendation system are some more areas where machine learning is used [5][10].

7.2 HISTORY OF MACHINE LEARNING

In today's era, Machine learning is not a new concept. Various industry pioneers have attempted to guide machine learning in the correct direction during the last 60 years. With his work at IBM, Arthur Samuel was involved in developing one of the earliest self-learning systems. He emphasised the games as a means of teaching the machines things for the first time. Frank Rosenblatt created the perceptron (the very first computing network) in 1957 to replicate the thinking pattern of the person's mind and demonstrated that only the perceptron can converge if what they were seeking to gain could be expressed. Minsky and Papert highlighted the preceptor disadvantage of neglecting the research functioning neurons for at least 10 years. Afterwards, in 1967, an approach known as the closest neighbour was developed to allow computers to recognise extremely simple patterns such as tracing a path through a city. Stanford University scholars created the 'Stanford Cart,' in 1979 which was a movable robot that can traverse obstructions in a house on its own.

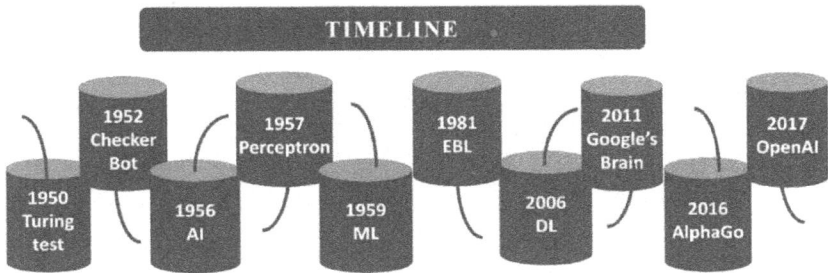

Figure 7.1 Machine learning timeline.

In 1981, Gerald Dejong pioneered the idea of Learning based on explanations, allowing a system to analyse and eliminate unnecessary material from supplied training data sets. Finally, in the 1990s, the machine learning approach became more data-driven rather than the traditional knowledge-based approach, in which programs were developed to analyse and draw conclusions from large amounts of data. In the last ten years, the launch of Google Brain in 2011, the introduction of Amazon's machine learning platform, and the establishment of a machine learning toolkit released by Microsoft in 2015, the realm of machine learning and artificial intelligence have witnessed several accomplishments and Google's artificial intelligence (AlphaGo) triumphed in the globe's utmost complicated Chinese board game versus Lee Sedol (a professional player), who won five times in a row in 2016. The development of a bot by OpenAI (created by Elon Musk) in 2017 that conquered the greatest player in the world in a professional eSport, Dota 2, is one of the cutting-edge successes in machine learning [8].

7.3 ARTIFICIAL INTELLIGENCE

Data mining has become a modernistic wave and a sub-field of Artificial Intelligence, similar to Machine Learning (ML). It is the mechanism of acquiring information from databases, sheets, and metadata, generating conclusions from all of this and translating it into a format that can be used for further processing. The training method involves data management, pre-processing, as well as graphical visualisation of the data [2]. Machine learning, a subcategory of artificial intelligence, is an algorithmic-based auto-learning mechanism that improves from its experiences. It employs a statistical learning approach that trains and develops on its own, without the need for human intervention. In a Deep Learning system, on the other hand, the algorithm learns from experience, from a huge database or prior data supplied as input to the model. Deep neural networks have multiple layers between their input and output whereas shallow neural networks have just two layers among their input and output. Artificial intelligence is a vast subject that generates humanoid robots, and the majority of AI study incorporates machine learning because intelligent behaviour necessitates a massive amount of data or insight [5]. The impact of AI may be observed in two ways: one, in deriving information from Big data in the domain of research, and secondly, in assisting doctors in healthcare [11].

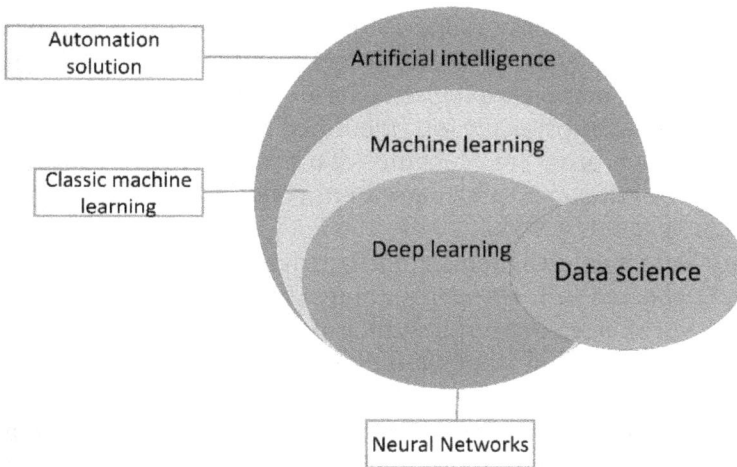

Figure 7.2 Artificial intelligence vs machine learning.

Despite the fact that AI has proven to be highly effective over the last 50 years, there are still many significant disputes in the area. Diverse study fields frequently do not interact, scientists employ various techniques, and there is currently no fundamental theory of intellect or learning that unifies the field [12]. Demonstrative AI specializes in addressing well-defined logical issues, but it frequently lacks problems requiring top-level pattern recognition, like picture categorization or speech recognition. Several more difficult jobs are there where ML and DL techniques excel [6]. Artificial intelligence may be regarded as computerized technologies that assist humans in doing time-taking operations. As a result, it reduces cost and time consuming and results in increased production. Deep learning is currently among the most prominent aspect. Deep Learning is an artificial intelligence discipline that includes the construction of complex computer simulations to solve commercial problems [2].

7.4 IMPORTANCE OF MACHINE LEARNING

There are several advantages of machine learning, and this is why machine learning research could not be ignored or disregarded. The following statements demonstrate the relevance of machine learning:

- Machine learning algorithms are frequently utilized to uncover relevant links and inferences which might be hidden within complex datasets. The quantity of knowledge accessible about specific jobs would be large for individuals to the explicit program. Machines that progressively absorb this information can catch a greater proportion of what people are expected to find.
- Machines designed by humans frequently fail to operate as expected in their surroundings. In reality, several aspects of workplace surrounding may not be completely understood at the time of model building. Machine learning techniques can be used to enhance system models.
- The surroundings evolve throughout time. Systems that can evolve to their surroundings would eliminate the need for continual redesigning. Humans are always learning new things regarding their tasks. In the world, there is a never-ending wave of new occurrences. It is impractical to continuously rework AI systems in order to incorporate new ideas and technologies [1].

7.5 TYPES OF MACHINE LEARNING

Machine learning entails discovering important regularities in the dataset and then applying that knowledge to make forecasts. This may be done in a variety of manners, leading to a variety of machine learning methods to select from. A standard categorization groups the various techniques based on learning type. Machine learning methods are divided into four categories based on this categorisation: supervised learning, semi-supervised learning, unsupervised learning, and reinforcement learning [3].

7.5.1 Supervised Learning

When a program learns, it gets both input and output data values and a supervised training system develops a mapping function that could identify the predicted outcome for the provided input values. The learning procedure is repeated till the algorithm achieves the desired level of precision [8]. The objective is to find the ideal method which best represents the correlation among the input and the target variable values using an algorithm. As this method knows what the output values will be, therefore this kind of learning is referred to as supervised learning. The method is trained on numerous instances and is given a response based on the accuracy of its forecast during the learning phase. The algorithm's evaluation is performed by comparing its forecasts to the dataset's real target parameter values. A supervised learning problem can be solved by using either a classification or a regression approach depending upon whether the dependent variable is continuous or categorized. For example, the computer might be made to learn to distinguish a spam email from a genuine email that is currently being utilised by Google to do spam filtering in Gmail. Some of the most prevalent supervised algorithms include K Nearest Neighbor, Linear Regression, Naive Bayes, Support Vector Machine, Random Forests and Neural Networks [3].

Let's take an example to understand clearly about supervised techniques:

Assume a property market wishes to forecast the cost of a house depending on its observable factors. To start, the organization would create a database with several samples. Each feature depicts a unique observation for house price and its related functions. The features are recorded, which are nothing more than attributes of a house that may be beneficial in forecasting values (e.g., overall square metres, area, total count of floors). The aim is to forecast the feature, which in this case is the property cost. Training, validation, and test data sets are created from the original dataset. Supervised learning utilizes models from the training data to relate characteristics to the output, allowing a program to forecast home prices on upcoming data. Such method is supervised as this model generates a method by finding the mapping function, F, that maps the features, X to the target variable, Y such that upcoming home values could be estimated utilising technique described by: $Y = F(X)$. The algorithm's performance is assessed using the test dataset, which contains information which the method hasn't perceived previously [6].

- **Regression** - The objective of a regression problem is to forecast a value on a scale that is continuous. Numerous problems that can be handled by a classifier can also be addressed by a regression algorithm (referred to as regressor), with the output being continuous [3].

- **Classification** - Classification algorithms seek to forecast collective identity, often referred to as labels as well as classes using a set of data. The popularity of this sort of algorithm might be influenced by the fact that the majority of clinical issues can be reduced to a classification decision [3].

7.5.2 Unsupervised Learning

Unsupervised learning is a process in which the computer is provided with a collection of unlabeled and unstructured training data and the unsupervised algorithm develops a function to uncover new characteristics in the supplied data [9]. No target value is available for this type of learning. The main objective is to uncover essential structures in data and recognise patterns in a database and categorise specific occurrences in the set of data [3]. The strategies are unsupervised as correlations that might or might not be present in a dataset are not revealed by a goal and must be determined by the algorithm [6]. Clustering and association issues may be addressed using unsupervised learning techniques. The K-means method for clustering and the apriori algorithm for association problems are two of the most widely used unsupervised learning algorithms. In the actual world, unsupervised methods have been utilised to divide insurance dataset networks into clusters for learning methods [13].

- **Clustering** - The unsupervised categorization of samples as groups is known as clustering (clusters). Cluster analysis finds a cluster center in data, where a cluster is a group of related data elements. A good clustering approach produces high-quality clusters with minimal inter-cluster resemblance and large intra-cluster resemblance. Homes, for example, can be classified into areas depending on their type, interior space and geographical region [14]. The cluster analysis has been used in brain disease studies for a variety of purposes, including examining the neurocognitive characteristics of bipolar illnesses [3].

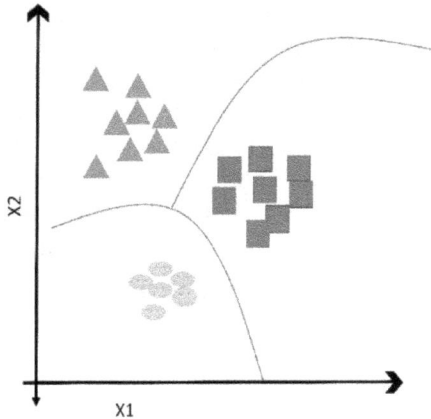

Figure 7.3 Clustering.

- **Association** - The finding of relationship connections or correlations among a set of components is referred to as an association. They are usually stated as a rule that indicates the feature value conditions which occur usually with each other in a particular data collection. Database tuples which satisfy X are able to satisfy Y, according to an association rule of the type which is expressed as $X \rightarrow Y$ and is interpreted as a tuple that says that X are likely to satisfy Y. Its analysis is offered with efficient algorithms, including apriori search, mining of multilevel and multidimensional associations, interval, categorical and numerical data mining associations, meta-driven or constraint-based models, and data correlations. In payment data analysis for marketing and advertising and other decision-making activities, association analysis is frequently utilised [14].

7.5.3 Semi-supervised Learning

Semi-administered learning can be considered as the fair compromise among regulated and solo learning and is especially helpful for the dataset that contains both labelled and unlabelled information (that is, all elements are accessible) [6]. This method is beneficial while accessing or measuring the target variable for all individuals when it is impracticable or too costly a measure. Longitudinal investigations of illness development, for example, that need many years of follow-up in order to acquire an accurate disease label, may be unfeasible or unaffordable. Under such circumstances, semi-supervised learning could be utilized to model both current labelled data (patients that have finished the involvement in the project) and unlabelled data (patients for whom a label has not yet been generated). For research with limited resources and time, this technique optimises the amount of input data [3]. It is frequently used in medical imaging, for which a clinician may tag a small selection of pictures and train a model with them. This algorithm can then be utilized to categorise the remaining unlabelled pictures in the set of data. The labelled data is utilised to develop a functioning model [6].

7.5.4 Reinforcement Learning

The system is subjected to the real world through which it makes judgments via observation and experimentation and adapts itself from its actions and improves from prior expertise in reinforcement learning. The computer gets positive feedback from the domain for each right choice, which functions as a reinforcement signal, and knowledge regarding rewarded state-action combination is recorded [9]. The objective of reinforcement learning, the same as operant conditioning, is to maintain a device that can adapt by engaging with the world. In reinforcement learning, the system is free to act, that is to find which behaviours maximise the compensation and minimise the punishment through trial and error [3]. In domains in which strategy formulation is critical to success, such as robotic automobiles (self-driving cars), reinforcement learning algorithms are utilised. Q-Learning and Markov Decision Processes are two of the most commonly utilised reinforcement algorithms [8]. Consider training an algorithm to operate the Super Mario Bros gameplay, in which the goal is to navigate the Mario figure out from the left of the screen to the right-hand side in terms of reaching the winning pole at the finish of every level while minimizing hazards such as opponents and traps. An algorithm would play games by itself in reinforcement learning. Many alternative control inputs could be attempted, and even the algorithm is compensated when Mario moves ahead (without receiving injury). The algorithm comes to discover what response is intended as a result of this procedure (e.g., travelling ahead is preferable to switching back, and leaping on attackers is preferable to striking them). Ultimately, the machine learns to traverse from beginning to end. While it has a great number of applications in the field of computer science and especially machine learning, it has yet to make a significant influence in clinical care [6].

7.6 MACHINE LEARNING ALGORITHMS

This part examines several machine learning algorithms, as well as their benefits and drawbacks.

7.6.1 Linear Regression

Linear regression is the most basic machine learning algorithm. The primary goal of regression analysis is to define a relation among one or more data variables and a single target objective. Linear regression is an analytical approach that uses a linear relationship to represent a collection of data to address a regression task. A slope-intercept form is given below and could be utilised to describe univariate linear regression, and that is a regression scenario in which only one feature can be used to forecast an output value.

$$y = Ax + B$$

A is a weight representing the slope, it defines how often a line grows on the vertical axis with each raise in x, and B is the interception, that specifies at which point, the line crosses the y - axis.

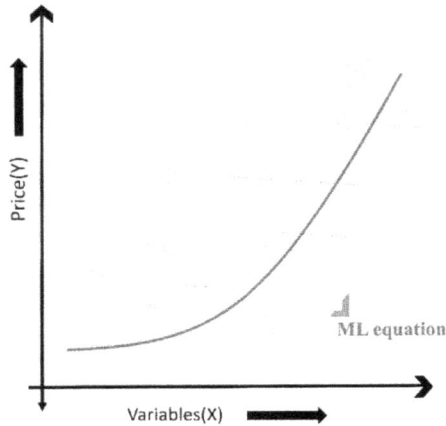

Figure 7.4 Linear regression.

Linear regression models the data by using a type of slope intersecting concept, in which the system's goal is to find the values of parameters a and b such that the identified line is best suited to connect the provided values of x with corresponding y values [6]. The goal is to forecast a result on a continuous scale [3]. Predicting the price of land, predicting future purchases, estimating academic tests, predicting variations in the price of stocks on the stock market, and so on are some instances of linear regression algorithm. In regression, the data sets are labelled, and the target variable's value is calculated by using values of the given parameters, making it a supervised machine learning method [9]. Multiple linear regression (MLR) is a technique that forecasts the outcome of a dependent variable by using many parameters in the model. It attempts to represent the linear connection between the experimental feature and target feature [15].

Advantages

- Linear regression is said to be a great fit if correlation among covariates independent variables and the target variable is confirmed to be linear.
- It is simple to grasp and avoid overfitting through regularisation.
- It shifts the focus away from statistical modelling towards data pre-processing and analysis.

Disadvantages

- As it oversimplifies real-life situations, it is not suggested for most practical situations.
- Whenever dealing with non-linear connections, this is not an acceptable method. It is tough to manage complicated models.
- It is challenging to correctly add the right polynomials to the system.
- When using ordinary least squares (OLS) to fit a regression line, the result is a line with a large residual sum of squares (RSS) rate.
- There will never be a linkage among the mean of the target and experimental features in actual problems that regression analysis predicts [9].

7.6.2 Logistic Regression

Logistic regression is a type of regression technique that uses one or more predictive variables to determine the result of a categorical feature [16]. Logistic regression is a classification method whose goal is to discover a link among a set of attributes and the likelihood of a specific result. It estimates the possibility of the class using a sigmoid curvature which is an S-shaped graph that transforms continuous or discrete numeric characteristics x to a single number value y in the range of 0 – 1 [16]. For instance, forecasting if a tumour is mild or severe, or if a mail could be categorised as junk mail or not, are illustrations of binomial logistic regression results. It may provide a multivariate result, such as predicting the favourite kind of food culture: Taiwanese, Estonian, Spanish, etc., or an arbitrary outcome, such as an online ranking from 1 to 5 and so on. As a result, logistic regression is concerned with the forecasting of a categorized objective variable. Estimating the chances of acquiring a certain ailment, like cancer detection, market analysis, spamming mail detection, and so on are some aspects of useful applications of regression analysis [16]. Logistic regression is mostly used for forecasting and also computing the possibility of success [17].

Advantages

- No transformation is necessary for input information, result-

ing in ease of implementation, computation speed and normalisation. This method is mostly utilised in the market for
issue resolution.

- Because the output of the logistic regression is a probabilis-
tic value, unique performance measurements must be specified to get a cut-off that can be used to execute the targeted
measured categorization.

- Small data noise and multicollinearity have no effect on logistic regression.

- The probabilities are restricted to a range of 0 to 1 [16].

Disadvantages

- If a non-linear issue cannot be addressed as its decision surface is deterministic and prone to overfitting, it would not
function well until all attributes are recognised [9].

7.6.3 Decision Tree

This algorithm is a supervised learning technique for classifying as
well as predicting situations. It is a visual graphic framework in
which internal nodes indicate tests on one or more characteristics
and external nodes reflect decision outcomes. A base, internal
nodes, edges, and external nodes make up the decision tree. The tree
is constructed from the leftmost side to the rightmost side, or from
the topmost root to the bottom terminals [16]. A tree starts with a
root, which would be the first point of choice for splitting the set of
data, generally includes a single characteristic that best separates
the information into its different groups. Every partition has such an
apex which links to either a fresh decision point with additional
characteristics to further split the data into homogenous groups or
an external node that forecasts the outcome. This array of steps
formulate recursive partitioning. It can be utilized to calculate the
expected usage of library resources and to solve detection difficul-
ties [6].

Advantages

- Ease of understanding and processing categorized and quantified information.

- Ability to fill in redundant data in characteristics with suitable and appropriate parameter values.
- Good efficiency owing to the effectiveness of the tree traversal method.

Disadvantages

- It could be unreliable and controlling the tree's growth might be challenging.
- It could discover the overfitting situation for which random forest is the solution that would be relied on an aggregation modelling technique, and
- It gives a regionally best solution rather than a universal solution [9].

A random tree is one randomly chosen from a range of available trees. It is called a random tree since every tree in the collection seems to have an equal chance of becoming randomly selected. These trees are very efficient since they yield to much more suitable model [16]. It is a decision tree based extension that operates with numerous decision trees. In this, rather than utilising each feature to build each decision tree, subsets of attributes are utilized to make a decision tree. The trees then forecast the consequence of a class, and the supermajority among the trees assists in obtaining model predictions. It is used to detect and categorize cataracts via ultrasound pictures, as well as to identify the people with glaucoma utilising information from the retinal nerve fibre layer and field of view [6].

7.6.4 Support Vector Machine

SVM is a learning model that examines data utilized for non-probabilistic grouping or the analysis of the regression problem [18]. It creates mathematical equations for multiple hyperplanes that may linearly give insight after displaying the training set over an n -dimensional grid, where n is the number of elements being assessed. If two or more components are examined, a hyperplane exists as a line, plane, or hyperplane. In the SVM method, the hyperplane with the largest margin is chosen for forecasting [19]. SVM aims to find a linear plane that has a large error margin in order to clearly differentiate the categories and decrease the sum of squares of errors. The goal of SVM is to find a linear plane that separates two or more classes and divides them with its greatest possible margin. A top-margin hyperplane is a type of SVM execution [2]. Cancer diagnosis, card fraud identification, handwriting identification, image recognition, language processing, and more applications are possible of SVM. It is useful when there are a high number of samples and characteristics [9].

Advantages

- When there are a lot of dimensions and a smaller number of training sets, this method is ideal.
- The separation margin is nearly flawless, resulting in a precise and important conclusion [2].
- SVM is simple to implement, and it gives a memory-efficient solution.
- It is capable of dealing with both organised and semi-structured data and is good for intensive tasks.
- There are lower chances of overfitting [9].

Disadvantages

- Convergence takes a long time with a big data set, therefore it may not be suitable for bigger datasets.
- SVM does not provide direct probable chances, these need to be calculated independently [2].
- With a big dataset, the model's performance suffers as learning time increases, making it difficult to find an acceptable kernel function.
- SVM performs poorly in the presence of noise and does not provide probability estimates, making the final SVM model difficult to comprehend [9].

7.6.5 Naïve Bayes

The Naive Bayes method is a simple technique that relies on conditional probability. In this technique, a probability table focusing on extracted features is used to evaluate the chances for predicting future observations and updating the training data. It makes the naive assumption that every input characteristic is not related to any other feature in any way, which is unlikely to be true [9]. When we do have two tasks P and Q, the conditional probability of P given Q is indicated by Bayes' rule as

$$P(B|A) = P(A|B)\frac{P(A)}{P(B)}$$

$P(A)$ and $P(B)$, respectively, are the probabilities of A and B, $P(A|B)$ is the probability of A given B, and $P(B|A)$ is the probability of B given A. The categorization is done using Bayes' theorem [2]. Naive Bayes could be utilized in areas like recommendation systems and prediction of cancer relapses or advancement after radiation therapy [9].

Advantages

- Simple to build and works well with limited training data.
- Scales linearly with the number of variables and pieces of information.
- Could deal with both numerical and categorical data.
- It is not delicate to inappropriate characteristics.

Disadvantages

- Properly trained and designed models mostly outperform it.
- Because there is no true variation for Naive Bayes, all data need to be preserved in order to re-educate the model. It will not measure if the count of labels is more than $100K$.
- When predicting, extra memory is a prerequisite at runtime than with logistic regression or support vector machine.
- It is operationally rigorous, especially for models with many variables [9].

7.6.6 K Nearest Neighbour

The KNN algorithm is a type of non-parametric algorithm that can be used for both regression and classification [18]. Whenever a new event is presented to the system, the method evaluates all information to track a subgroup of instances that would be most similar to it, and then use that to forecast the results. This method has two driving factors: the closest count of samples that utilise k and a measure to quantify what is intended by the nearest thing. Every time we apply the k-NN method, we must give positive numeric values for k. While computing a new case, this determines how many previous cases are evaluated [14]. The KNN technique is used for finding the shortest path by employing the Euclidean measure. The below equation is used to calculate the Distance measure $D(a, b)$ between 2 points a and b:

$$D(a,b) = \sum_{i=1}^{M} \sqrt{a^2 - b^2}$$

where M is the number of features such as $a = \{a_1, a_2, a_3, \dots a_n\}$ and $b = \{b_1, b_2, b_3, \dots b_n\}$ [20]. KNN could be utilized in recommender models, to diagnose several illnesses with indicators, to identify writing, to do research prior to authorising loans, to recognise videos and photos etc. [9].

Advantages

- This approach is straightforward and easy to set up.
- The model's development is low-cost, and it's a versatile categorization method.
- It's perfect for multifunctional classes.

Disadvantages

- Necessary to measure the distance between the k nearest neighbours.
- The method becomes more costly in terms of time as the training set grows larger.
- Precision is harmed by the elements that are unnecessary or unrefined.
- Lazy learner as it calculates the distance over k neighbors and never generalize the training features and possess them entirely.
- Handling huge data sets leads to costly calculations. Greater dimensions of data result in the reduction of accuracy [9].

7.6.7 K-Means clustering algorithm

This method is essentially a vector quantization in which it is necessary to define a certain number of clusters. Let us assume we have 2000 items and we want to frame 4 sets then N (items) = 1000 and S (sets) = 4. All of these groups have a centroid, a point from where the distance of the entities will be evaluated. It is used to identify irregularities and rapid computations [21]. The clusters are specified by repeating the process over the distances of the entities to be calculated that is closer to the centroid [7].

Advantages

- When there are many variables, it is more effective than hierarchical grouping.
- It produces clusters that are narrower than hierarchical clusters with small k and globular clusters.
- The attractiveness of this method is its simplicity in order to facilitate the employment and understanding of clustering results.
- It is competent from a computational standpoint since the order of complexity is $O(N \times S \times d)$

Disadvantages

- The k value is hard to measure.
- When groups are globular, performance suffers, and distinct starting partitions result in different end clusters, disrupting efficiency.
- When the cluster size and density of the input data vary, performance suffers [9].

7.7 HYPER PARAMETER TUNING

At the time of model building process, we must iterate over multiple factors that are used for effective weight calculations in our model; these parameters are known as hyper parameters. Hyper parameter tuning is done to achieve the most accurate results. A network can be trained on a lot of concepts by analysing the training dataset, though there are some characteristics which must be supplied. Such variables determine a network's topology, procedure, and learning rate etc [2]. The Major aim of the machine learning procedure is to increase the performance score which can be done by getting an ideal combination of factors that results in obtaining the best accuracy for the model. The superlative approach to selecting optimal hyper parameter values is done by trial and error of all likely arrangements of parameter values. Scikit-learn offers GridSearchCV as well as RandomSearchCV methods to enable the involuntary tuning of hyper parameters. In GridSearchCV, we may specify a set of parameter values for the model, and models are created for all feasible solutions of a supplied range of hyper parameter values, with the accurate estimation chosen on the basis of cross-validation score.

Drawbacks with GridSearchCV

- Calculating the supplementary parameter values using the Grid Search method is computationally expensive.
- Not able to provide nearly optimal parameters.

The Random Search algorithm tests for random combinations of a range of given parameter values [22]. In other words, Random Search is a method where arbitrary arrangements of the hyper parameters are used to find the optimal values to train the model [17]. Altering hyper parameters is commonly defined as a measure of the model selection process. Making a prosperous model often consist of testing various methodologies. Model selection is a practice of selecting between various models on the basis of performance with different algorithms, feature sets, or selecting among different hyper parameters [23].

7.8 MODEL TRAINING PROCESS

Problem formulation

Problem formulation is the most basic and initial step in which a problem is defined along with the selection of a suitable defined feature set, task and target variable. The target variable is also called a dependent variable which an algorithm is going to predict. The feature set is also referred to as independent variables used as input to predict the target variable. The feature vector is defined as all observations as a single representation. The task is a function that an algorithm performs on a feature and the target variable. All these components help to formulate a problem on which the machine learning algorithm works. It is an important step that let us know what we are going to predict [23].

Data preparation

Real-world data is messy, unclean and faulty, so there is a need to explore, analyse and clean the dataset so that the quality of the available data can be improved which results in better and accurate precision. In the dataset, mostly it can possess issues such as duplicates, categorical variables, missing data and imbalanced dataset, outliers etc. Duplicate data should be removed from the dataset as it can degrade the accuracy of the model. Categorical data needs to be converted to numeric data as models use numeric values for calculations and it can be done using dummy variables or one hot encoder method. Missing data can contain NULL, NAN values which can be handled by deleting those rows or mean median or mode imputation. Outliers may or may not be dangerous for the model, it depends upon the type of data available [2].

Feature engineering

EDA (exploratory data analysis) can be univariate or bivariate and used to recognize size, patterns and visualize the features [2]. Feature engineering is defined as operations performed on raw data to extract useful features out of the dataset and often uses feature engineering methods such as feature selection, feature extraction, feature scaling/normalization and dimensionality reduction [23]. Feature selection is described as a procedure in which irrelevant characteristics are removed while the remaining features are kept. The creation of new features from existing ones is done in feature extraction, which is primarily helpful in the case of correlated data and leads to the reduction in the dimensionality via Principal component analysis (PCA) and Independent component analysis (ICA) [24]. The majority of the time, data should be of the same scale so that feature scaling and feature normalisation may be performed on the data.

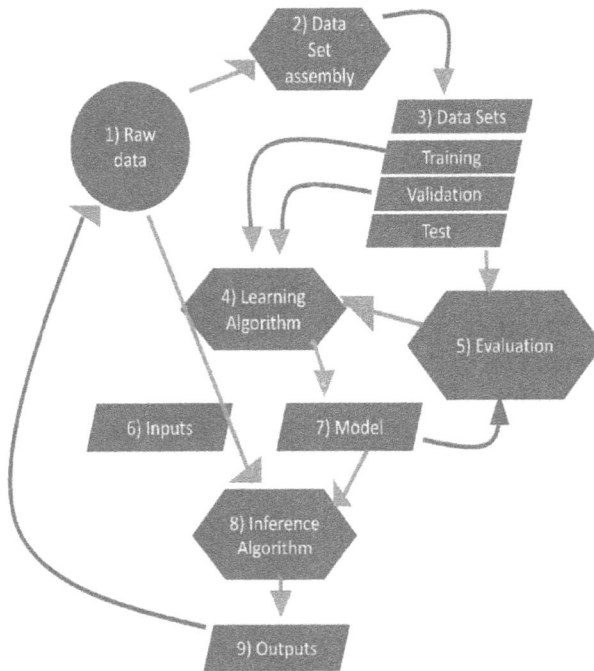

Figure 7.5 Model training and evaluation process.

Model training

Model training is a process in which the model is trained using a suitable machine learning algorithm by dividing the dataset into training and test sets. First of all, an algorithm is trained using a training set and the model is evaluated using a test set where the performance of the model is checked on unknown data. Sometimes for estimating the predictions more precisely, cross-validation is used which splits the data into training and testing many times and results in the output by considering the prediction of all cases. It leads to increased stability of the model [23] [2].

Evaluation

Some techniques are recommended for model performance evaluation, and the metrics considered for evaluation include correlation coefficient (R), mean absolute error (MAE) and Root mean square error (RMSE) etc [25]. Part from these True Positive rate (TP), False Positive rate (FP), True Negative (TN), False Negative (FN) rates are the other metrics used for evaluation, and these parameters can be represented in the form of a Confusion Matrix, which can be used to calculate performance, accuracy, sensitivity, specificity, and area under the receiver operating characteristic (AUC-ROC). By evaluating all these factors, we can come to know how well the model has been trained [23]. The receiver operating characteristic (ROC) chart states the relationship amongst sensitivity and 1 - Specificity. The nearer the ROC curve is to the upper left corner, the well-trained model will be. The area under the curve (AUC) value always lies in the middle of 0 and 1 and is a good indicator of model performance [26]. Testing a dataset on an out-of-time dataset is always a smart idea [6].

7.9 IDENTIFYING POTENTIAL CUSTOMERS

In this section, the problem of identifying potential customers who can buy a certain product has been analyzed using two of the most important machine learning algorithms. The algorithms that have been used in this work are Logistic Regression and KNN (K-Nearest Neighbor) where Logistic Regression is a supervised learning algorithm and KNN is a clustering technique. Both the algorithms have been applied to solve the classification problem. Not only this section discusses the algorithms used but also contains a detailed explanation about the model creation pipeline. The data has been intentionally chosen to be the same for both the techniques in order to provide a comprehensive discussion and analysis about the techniques.

Pseudocode of the algorithms (Model creation pipeline):

- Import libraries.
- Load the data.
- Split the data into train, validation and test datasets.
- Enable K-Fold cross-validation for an accurate evaluation of the performance.
- Standardize the data for faster convergence of the model.
- Train the model.

- Make the predictions for the classification problem for the validation set.
- Evaluate the performance in terms of accuracy for the validation set.
- Make the predictions for the test set.
- Evaluate the performance in terms of accuracy for the test set.
- Perform visual analysis of the results.

Logistic regression

Logistic regression algorithm is a supervised learning algorithm where the main emphasis is on finding the optimal values of W and b from the following equation

$$Z = Wx + b$$

where W is the weight and b is the intercept. For the classification problem, y_pred is calculated using the sigmoid function $h(z) = \frac{1}{1+e^{-z}}$ such that $y_{pred} = \begin{cases} 1 \ if \ h(z) \geq 0.5 \\ 0 \ if \ h(z) < 0.5 \end{cases}$. The sigmoid function is shown in Figure 7.6.

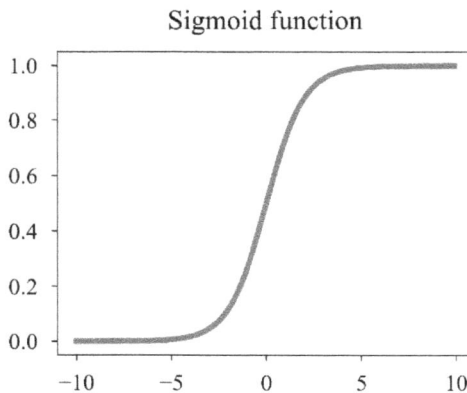

Figure 7.6 Sigmoid function.

The algorithm works on the concept that the loss/error should be minimum and this error is calculated using the cost function $f(n)$ as shown below;

$$f(n) = \frac{-1}{N} \sum y * \log y_{pred} + (1 - y) * \log(1 - y_{pred})$$

Let us now apply logistic regression to the data for classifying the data. The dataset [27] used for this work contains two columns 'Age' and 'Salary' as the input data and an output column that contains binary values 0 and 1 which signifies that whether the person with certain age and salary will buy the product or not. The data is loaded and split into train, validation and test sets where the train set contains 256 records, validation contains 64 records and the test set contains 80 records. Performance evaluation of the model is done on the validation set using the cross-validation technique to get a more accurate estimation of the performance of the model. Prior to training the model, the data is standardized so as to make the convergence faster. Once the model is fully trained and evaluated for performance, the model is then applied to the test dataset to find out how it performs on the new and unseen data. In this work, StratifiedKFold cross-validation with $K = 5$ has been used to train and evaluate the model on 5 different sets as it provides the model with a good variety of different types of data. The results show that the Logistic Regression model is able to accurately classify the data with a mean accuracy of 82% on the validation set and an accuracy of 92.5% on the test set. Visual classification of the data can be seen in Figure 7.7.

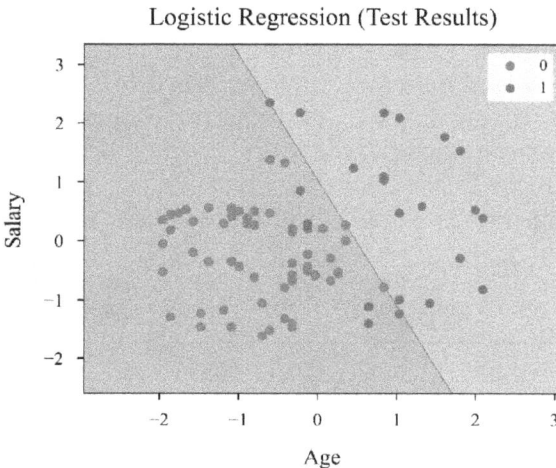

Figure 7.7 Results obtained by logistic regression.

In the figure, it can be observed that there are very few misclassifications and that too can be adjusted using the hyperparameter tuning process.

K-Nearest Neighbours

K-Nearest Neighbours (KNN) is an unsupervised learning algorithm where the main emphasis is on making clusters with maximum intra-cohesion and minimum inter-coupling. The KNN algorithm works on the basis of the distance between different data points where the distance can be either euclidean distance, manhattan distance, or minkowski distance. All those data points that have minimum distance among themselves are grouped together to create a cluster and similarly, various other clusters are also created such that the distance between different clusters should be maximum. The number of neighbours that can be chosen is a very subjective question and generally, the optimal number of neighbours that gives the maximum accuracy is around $\sqrt{n}$, where n is the number of data points but this is not a fixed criterion. In this work, there are 400 records so $n = 400$ and the number of neighbours that can be chosen is 20 but it was found in the study that the $5 \leq$ neighbours ≤ 20 gives the same performance in terms of accuracy and therefore in this work number of neighbours chosen is 5. Again the KNN algorithm was applied to the same dataset on which the logistic regression was applied to compare the performance of the two algorithms and it was observed that the KNN algorithm obtains a mean accuracy of 89.68% on the validation set and an accuracy of 95% on the test set which is significantly higher than the Logistic Regression model on the given dataset. A comparative analysis of both techniques has been shown in Table 7.1.

Table 7.1 Comparative analysis of logistic regression and KNN

Accuracy (Logistic Regression)	Accuracy (KNN)
0.84375	0.90625
0.76562	0.9375
0.875	0.89062
0.78125	0.85937
0.8125	0.89062
0.81562	**0.896872**

On the other hand, KNN algorithm being a clustering technique took around 10.62 seconds to compute the predictions in comparison to the Logistic Regression model which took almost 0.30 seconds. Visual classification of the data using KNN clustering technique can be seen in Figure 7.8.

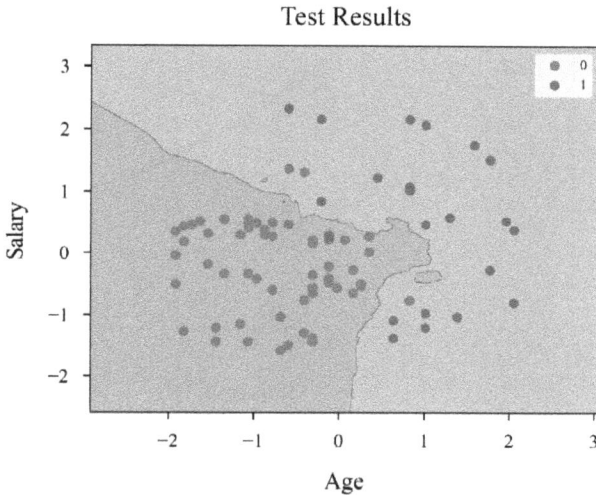

Figure 7.8 Results obtained using KNN technique (StratifiedKFold).

Again, it can be observed from the figure that there are very few misclassifications which can be reduced by hyperparameter tuning. A sincere effort was made to assess the performance of the KNN algorithm using both KFold cross-validation and StratifiedKFold cross-validation and it was found that KFold cross-validation produces extreme minimum and maximum accuracy values in comparison to the StratifiedKFold cross-validation as shown in Figure 7.9. An analysis of this phenomenon was done, and it was concluded that KFold cross-validation randomly picks up the data to create various folds whereas StratifiedKFold cross-validation makes a careful selection of the data based on the percentage of each class present in the original data.

KFold vs StratifiedKFold

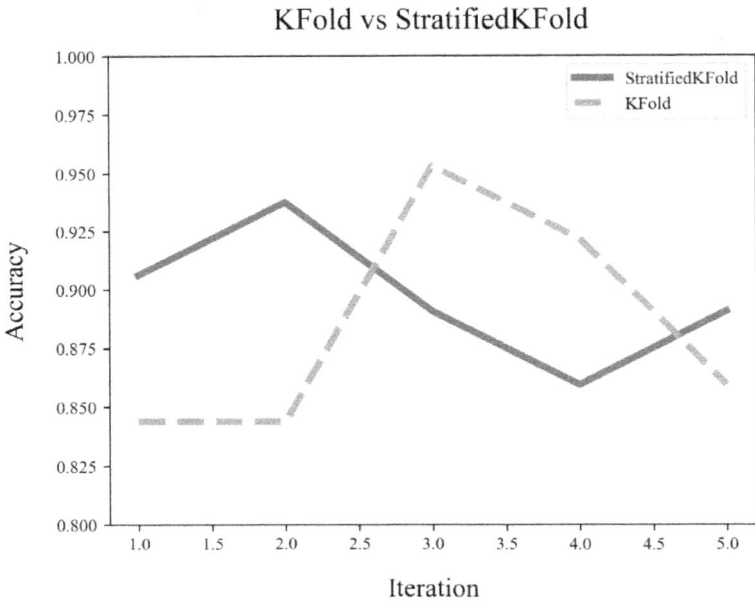

Figure 7.9 Comparison between KFold and StratifiedKFold.

Visual representation of the KNN technique using KFold rather than
StratifiedKFold is shown in Figure 7.10.

Test Results

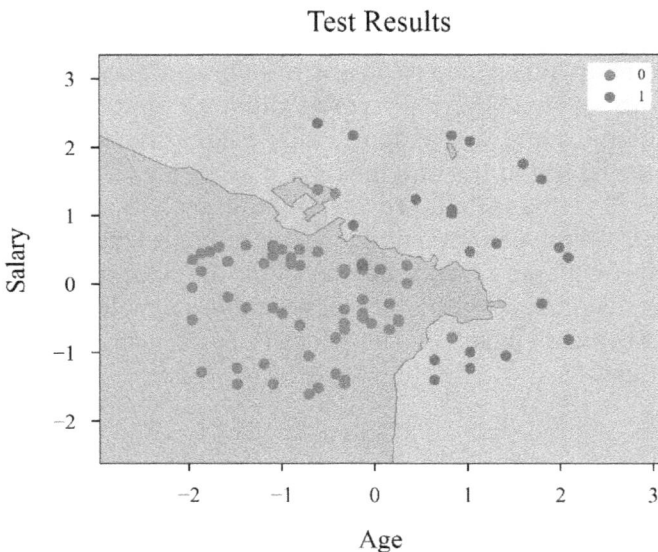

Figure 7.10 Results obtained using KNN technique (KFold).

It can be observed from Figure 7.10 that using KFold rather than StratifiedKFold results in the creation of different contours.

7.10 CHALLENGES IN MACHINE LEARNING

Data sources may possibly not be reliable, appropriate and consistent. In machine learning, the model learns everything from data, therefore it is necessary to use suitable data and make sure that it should not suffer from any type of data poisoning. The major necessity is to guard delicate and private data which is used while training the model. Sometimes model learns from the training data set more exhaustively and as a result, fails to generalize on the test set which thus results in the problem of overfitting [28]. Algorithms are data sensitive as functions and logic are learned from the training set so training data should be good enough. ML applications test cases often fail due to incorrect behaviour. A vast amount of time and money are needed in model creation. It is always difficult to evaluate all the parameters of the test case are another major challenge [29]. The model may experience a black box problem that happens when the testing data differs greatly from the dataset on which the model was trained. Traditional machine learning algorithms are restricted in their capacity to route natural data in its raw state. For years, building a pattern-recognition or machine-learning system required careful engineering and extensive domain knowledge to devise a feature extractor that can transform raw data into a suitable feature vector from which the training component could recognise or categorise regularities in the input [30].

7.11 CONCLUSION

Machine learning is being accepted and used in a wide variety of technological applications. It is a new field of artificial intelligence that will lead the way forward in the coming years. In this chapter, an attempt has been made to touch its fundamentals concepts along with different approaches and algorithms. A brief overview of the model training process, applications, challenges and hyper parameter tuning has also been provided. To reduce the danger of failed projects, it is critical to have a thorough grasp of the training process and the problems that are coming up in the field of machine learning. With a superficial understanding of the algorithms, one may select appropriate approaches to use to get better outcomes.

References

1. Tiwari, A. K. (2020) Introduction to Machine Learning. In: *Deep Learning and Neural Networks*, IGI Global, pp. 41–51.
2. Verdhan, V. (2020) *Supervised Learning with Python,* CA: Apress, Berkeley.
3. Vieira, S., Lopez Pinaya, H., and Mechelli, A. (2020) Introduction to machine learning. *Machine Learning: Methods and Applications to Brain Disorders*, 1-20.
4. Lin, J., and Kolcz, A. (2012) Large-scale machine learning at twitter. Proceedings of the international conference on Management of Data - SIGMOD '12. Online: https://dl.acm.org/doi/10.1145/2213836.2213958 (accessed on 20th August 2021).
5. Sharma, N., Sharma, R., and Jindal, N. (2021) Machine Learning and Deep Learning Applications-A Vision. *Glob. Transitions Proc.,* 24-28.
6. Choi, R. Y., Coyner, S., Kalpathy-Cramer, J., Chiang, M. F., and Peter Campbell, J. (2020) Introduction to machine learning, neural networks, and deep learning. *Transl. Vis. Sci. Technol.,* 9(2), 1-12.
7. Nath, V., and Levinson, E. (2014) Machine Learning. In: *Springer-Briefs in Computer Science,* pp. 39–45.
8. Patel, L., and Gaurav, K. A. (2020) Introduction to Machine Learning and Its Application. In: *Applications of Artificial Intelligence in Electrical Engineering,* IGI Global, pp. 262–290.
9. Ray, S. (2019) A Quick Review of Machine Learning Algorithms. Proceedings of the International Conference on Machine Learning, Big Data, Cloud and Parallel Computing: Trends, Perspectives and Prospects. Online: https://ieeexplore.ieee.org/document/8862451 (accessed on 20th August 2021).
10. Parker, M. (2017) Introduction to Machine Learning. *Digital Signal Processing 101,* Elsevier, pp. 347-359.
11. Jones, L. D., Golan, D., Hanna, S. A. and Ramachandran, M. (2018) Artificial intelligence, machine learning and the evolution of healthcare. *Bone Joint Res.,* pp. 223-225.
12. Moor, J. (2006) Artificial Intelligence Conference: The Next Fifty Years. *AI Mag.,* pp. 87–91.
13. Li, S.-H., Yen, D. C., Lu, W.-H., and Wang, C. (2012) Identifying the signs of fraudulent accounts using data mining techniques. *Comput. Human Behav,* 28(3), pp. 1002–1013.
14. Zhang, C., and Zhang, S. (2002) Association Rule Mining. *LNAI,* Springer, USA.
15. Josephine, B. M., Ramya, K. R., Rao, K., Kuchibhotla, S., and Rahamathulla, S. (2020) Crop Yield Prediction using Machine Learn-

ing. *ADALYA J.*, 9(2), pp. 2102–2106.

16. Chaudhary, A., Kolhe, S., and Kamal, R. (2013) Machine Learning Classification Techniques: A Comparative Study. *Int. J. Adv. Comput. Theory Eng. Adv. Comput. Theory Eng.*, 2(4), pp. 2319–2526.

17. Saw, M., Saxena, T., Kaithwas, S., Yadav, R., and Lal, N. (2020) Estimation of Prediction for Getting Heart Disease Using Logistic Regression Model of Machine Learning. Proceedings of International Conference on Computer Communication and Informatics (ICCCI). Online: https://ieeexplore.ieee.org/document/9104210 (accessed on 20th August 2021).

18. Zhang, S., Wang, S., and Habetler, T. G. (2020) Deep Learning Algorithms for Bearing Fault Diagnostics - A Comprehensive Review. *IEEE Access,* pp. 29857–29881.

19. Banerjee, R., Bourla, G., Chen, S., Kashyap, M., and Purohit, S. (2018). Proceedings of IEEE MIT Undergraduate Research Technology Conference (URTC). Online: https://ieeexplore.ieee.org/document/9244782 (accessed on 20th August 2021).

20. Neagu, D. C., Guo, G., Trundle, P. R., and Cronin, M. A Comparative Study of Machine Learning Algorithms Applied to Predictive Toxicology Data Mining. *Altern. to Lab. Anim.* 35(1), pp. 25–32.

21. Pelleg, D., and Moore, A. (2000) X-means: Extending K-means with Efficient Estimation of the Number of Clusters. Proceedings of the 17th International Conf. on Machine Learning. Online: http://www.aladdin.cs.cmu.edu/papers/pdfs/y2000/xmeans.pdf (accessed on 20th August 2021).

22. Swamynathan, M. (2019) *Mastering Machine Learning with Python in Six Steps.* Berkeley, CA: Apress.

23. Vieira, S., Lopez Pinaya, W., Mechelli, A., (2019) Main concepts in machine learning. *Machine Learning: Methods and Applications to Brain Disorders,* Elsevier, pp. 21–44.

24. Alpaydin, E. (2011) Machine learning. *Interdiscip. Rev. Comput. Stat.,* Wiley, 3(3), 195–203.

25. Olyaie, E., Zare Abyaneh, H., and Danandeh Mehr, A. (2017) A comparative analysis among computational intelligence techniques for dissolved oxygen prediction in Delaware River. *Geosci. Front.,* 8(3), pp. 517–527.

26. Hirshberg, G. (2017) Over. *The Doubled Life of Dietrich Bonhoeffer,* Lutterworth Press, 59(3), pp. 411–413.

27. https://www.kaggle.com/rakeshrau/social-network-ads (accessed on 20th August 2021).

28. Mcgraw, G., Bonett, R., Shepardson, V., Figueroa, H. (2020) The Top 10 Risks of Machine Learning Security. *Computer (Long. Beach. Calif.),* 53(6), pp. 57–61.

29. Huang, S., Liu, E. H., Hui, Z. W., Tang, S. Q., and Zhang, S. J. (2018)

Challenges of testing machine learning applications. *Int. J. Performability Eng.*, 14(6), 1275–1282.

30. Lecun, Y., Bengio, Y., and Hinton, G. (2015) Deep learning. *Nature*, 521(7553). pp. 436–444.

CHAPTER 8

LATEST TRENDS IN PRINTED ANTENNA DESIGNS FOR EMERGING WIRELESS COMMUNICATION SYSTEMS

Smriti Agarwal and Anand Sharma
Department of Electronics and Communication Engineering
Motilal Nehru National Institute of Technology (MNNIT) Allahabad,
Prayagraj, India

8.1 INTRODUCTION

There has been an unprecedented advancement and rapid techno-logical changes in the field of wireless communications with the emergence of applications like 5G, internet of things, wireless sens-ing, satellite communication, body area network and automation in different fields. Hence, the related antenna technology has been con-tinuously improving and expanding in order to meet the stringent requirements of faster data transfer speeds, dense cellular network-ing, broader range implementations and compactness, which are the indispensable criteria for futuristic wireless communication sys-tems [1, 2]. The induction of novel, emerging wireless communica-tions systems with integrated applications make the antenna design a challenging task which not only demands multiple antennas to op-erate at new bands but also the antenna features like; multi-band/ultra-wideband operation, adaptability to diverse scenarios, small size for integration and easy transportability of the handheld devices [3-5].

The printed antennas have become a preferred choice because of their competitive features like, low profile, conformal, lightweight, less cost and easy integration which are much needed for compact and efficient wireless communication [6]. Printed antennas are not only used as communication element rather it is also useful in appli-cations like wireless sensing, IoT, imaging, automotive collision avoidance and bio-medical applications [3-13]. This chapter is broadly divided into two categories. Firstly, the performance pa-rameters of printed antennas are discussed in detail that is required for a good wireless communication system design viz., radiation pat-tern, impedance bandwidth, antenna gain, polarization, MIMO diversity performance parameters. A summary of different performance parameters and physical antenna parameters is shown in Figure 8.1. These parameters are needed to be optimized in order to make antennas more compliant to diverse wireless communication scenarios supporting manifold frequency bands, reconfigurability, beam steering, size miniaturization and easy integration etc. The second portion deals with the emerging applications and developments in printed antennas in the domain of the internet of things (IoT), 5G mobile communication, satellite communications, vehicular communication, near field wireless sensing and imaging.

Figure 8.1 Demonstration of different antenna design challenges for wireless communication system applications.

8.2 ANTENNA PERFORMANCE ATTRIBUTES

8.2.1 Radiation Pattern

Any practical RF antenna does not radiate power equally in all directions rather it will radiate more power in some directions as compared to others. The actual power distribution is highly dependent upon the antenna type, its design, size, feed, wave polarization and a variety of other environmental factors. The directional pattern ensures that the power radiated is focused in the desired directions

[14]. Thus, an important concern of antenna performance is the radiation pattern that determines its radiation characteristics like beam shape, main lobe beamwidth, sidelobe level, directivity, front to back ratio, polarization and power radiated. A typical 3D antenna radiation pattern is shown in Figure 8.2. The main beam represents the maximum radiation in the desired direction. Minor lobes represent radiation in the undesired direction, and it is aimed to be minimized preferably below -30dB. Beamwidth is the angular separation amongst the two points where power pattern magnitude is half of the maximum value, also known as half-power beamwidth. Beamwidth decreases with an increase in substrate height and patch width. The radiation pattern is although the three-dimensional plot but for most practical applications it is measured in particular planes viz. E plane and H plane to infer the full needed information of any antenna. For printed radiators, surface waves due to substrate permittivity, feed line radiation and edge effects are the few parameters to be optimized for good pattern control with low sidelobe and cross-polarization levels. For wireless sensing radar applications, beamwidth determines resolution capability i.e. the two objects should be placed at minimum angular separation equal to HPBW (~FNBW/2) in order to be uniquely identified [15, 16]. Thus, another challenge for the impending printed antenna array technology is to reduce down the extraneous radiation levels of the side lobes.

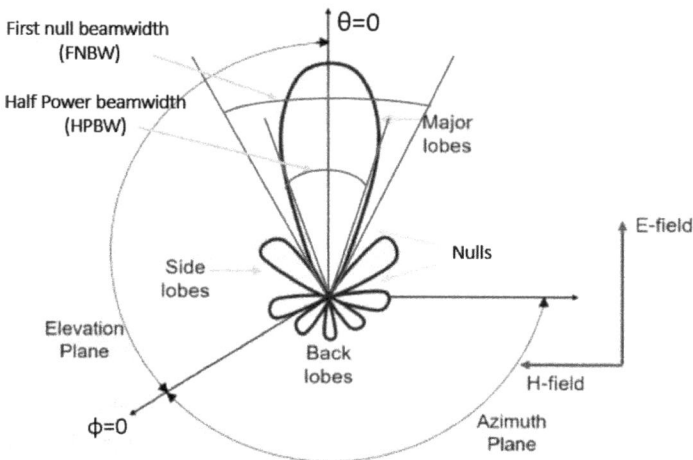

Figure 8.2 Demonstration of antenna radiation pattern, beamwidth and azimuth/elevation planes.

8.2.2 Impedance Bandwidth

An RF antenna is like a tuned circuit having a certain resonant frequency at which capacitive and inductive reactance cancels. Thus, there is only a limited bandwidth over which an RF antenna design can operate efficiently. Impedance bandwidth is a parameter that defines the frequency range over which antenna impedance is matched to the feedline [17]. A radio transmitter may severely damage if the antenna is operated outside its operating range, similarly to ensure the best reception, antenna design should be tuned within the optimum bandwidth limits only. The bandwidth (BW) is usually quantified as the frequency span for which voltage standing wave ratio (VSWR) value is less than two (alternatively equivalent return loss is 9.5 dB or 11% reflected power) [14]. VSWR is a function of reflection coefficient (Γ) which is a measure of reflected power at the antenna feed and reflection coefficient in terms of antenna input impedance (Z_{in}) and feed line characteristic impedance (Z_0) is defined as shown in equation (1), (2) [18]. Away from resonance, the input impedance will be mismatched, creating a large reflection and higher VSWR of (S >2).

$$VSWR = S = \frac{1+|\Gamma|}{1-|\Gamma|} \quad \text{where} \quad \Gamma = \frac{Z_{in}-Z_o}{Z_{in}+Z_o} \tag{1}$$

$$BW = \frac{(S-1)}{Q\sqrt{S}} \, 100\% \tag{2}$$

Thick and/or lower-permittivity substrate reduces quality factor (Q) and hence increases bandwidth but results in loss of radiated power due to trapped surface waves. Hence, various bandwidth extension schemes have been researched without reducing down the overall antenna performance for good communication. For instance, the inclusion of parasitic patches, introducing multiple resonances, stacking or multilayering, use of impedance matching networks etc are the few commonly used techniques.

8.2.3 Antenna Gain

Gain is also an important parameter that signifies the energy radiated by the antenna in the given direction as compared to the reference antenna with the same power input [14]. In order to improve system performance and for efficient, long-distance and reliable wireless communication link design, antenna gain is to be maximized. It is measured in decibels (dB). Ideally, the reference antenna is an isotropic antenna (gain in dBi) but to obtain a more practical, realizable gain prediction a dipole antenna (gain in dBd) is used as the reference antenna. Basically, gain (G) and directivity (D) are the same however, gain includes radiation efficiency (e_{cd}) of the antenna and takes into account losses within the antenna (conductor loss (e_c), dielectric loss (e_d) and reflection loss etc.):

$$\text{Gain } G(\theta, \emptyset) = e_{cd}D(\theta, \emptyset) \tag{3}$$

where, $e_{cd} = e_c e_d; \ 0 < e_{cd} < 1$

The Friis transmission equation (4), suggests that for ensuring high received power at a distance R the gain of radiating antenna should be high [14].

$$\frac{P_r}{P_t} = \left(\frac{\lambda}{4\pi R}\right)^2 G_t G_r \tag{4}$$

where, $\left(\frac{\lambda}{4\pi R}\right)$ is the free space loss factor, that signifies the spreading of radiating energy w.r.t. distance in free space. Antenna gain can be increased by increasing substrate height and patch width or by using a larger aperture antenna which could capture more energy from the incident EM wave [14]:

$$A_e = e_{cd} \left(\frac{\lambda^2}{4\pi}\right) D_0 \tag{5}$$

There is a trade-off situation on increasing the gain that beamwidth will also be reduced, hence, this more directional antenna will be critical to orient in the desired direction for proper reception of the power. Therefore, careful design and choice of optimal parameters are critical for a good communication radio link setup.

8.2.4 Polarization

Polarization is a very important parameter for antenna designing. In a simple language, polarization can be defined as locus traced by the tip of the E-field vector in the variation of time at the given point in space [14]. It can be classified into three categories: linear, circular and elliptical as shown in Figure 8.3. In the case of linear polarization, the tip of the E-field is always directed along the line. On the other hand, in the case of elliptical and circular polarization, the tip of the E-field dashes an ellipse and circle respectively.

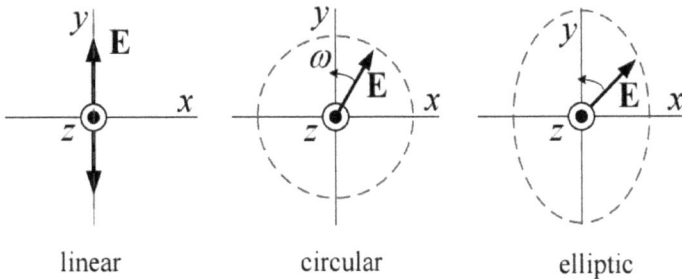

linear   circular   elliptic

Figure 8.3 Locus traced by the tip of E-field vector in different categories of polarization.

In the elliptical polarization, two field components are orthogonal to each other, but their amplitudes are not equal ($E_X \neq E_Y$; $\Delta\emptyset = 90^0$). Circular and linear polarization is the special case of elliptical polarization ($E_X = E_Y$; $\Delta\emptyset = 90^0$). In circular polarization, two orthogonal components are equal in amplitude. While in linear polarization, the phase difference between two fields components is in time phase or 180^0 out of the phase ($\Delta\emptyset = 180^0$) [14]. Out of these three types of polarization, circular polarization is a very important one in the current scenario. It is due to its capability of making the transmitting and receiving antenna orientation independent. This feature makes the circularly polarized antennas widely suitable for satellite applications as well as an efficient receiving antenna in a fading environment. Depending upon the rotation of E-field, circular polarization can be again classified into two categories i.e. right-handed circular polarization (RHCP) and left-handed circular polarization (LHCP). If the revolution of the E-field is clockwise with respect to the observer, then it is known as RHCP. Similarly, the revolution of E-field in an anti-clockwise direction with respect to the observer is considered as LHCP.

For identifying the type of polarization during the antenna design on the EM simulator, axial ratio (AR) is a very important parameter. It is the ratio of the major to the minor axis of the ellipse as shown in equation 6 [14].

$$\text{Axial Ratio (AR)} = \frac{\text{Major Axis Length}}{\text{Minor Axis Length}} = 20 \log\left(\frac{E_{max}}{E_{min}}\right) \text{dB} \qquad (6)$$

In the case of circular polarization, major and minor axis length is approximately the same. So, in an ideal case, the value of AR is one (in linear scale). But, for practical cases, it is difficult to achieve the unity AR or 0 dB AR value. So, the CP frequency range is considered, if the value of the axial ratio (AR) is just below 3 dB.

8.2.5 Diversity Parameters

Designing of efficient MIMO radiator is identified by the diversity parameters such as Envelop Correlation Coefficient (ECC), Diversity Gain (DG), Mean Effective Gain (MEG) as well as Channel Capacity Loss (CCL). The first and very important diversity parameter is ECC. It tells us the measure of correlation among the different antenna ports placed on the same substrate [19]. The value of ECC can be calculated in two different ways *i.e.*, S-parameter and far-field parameter. The formula of ECC as a function of S-parameter is defined as shown in equation (7) [19]:

$$\text{ECC}_S = \frac{|S_{11}^* S_{12} + S_{21}^* S_{22}|^2}{\left(\left(1 - (|S_{11}^2| + |S_{21}^2|)\right)\left(1 - (|S_{22}^2| + |S_{12}^2|)\right)\right)} \qquad (7)$$

where S_{11} and S_{22} is the reflection coefficient value at port-1 and port-2 respectively. Similarly, S_{12} and S_{21} is the isolation level between port-1 and port-2. Here, equation (7) holds true only for the two-port MIMO antenna system. Likewise, ECC can also be evaluated using the formula as given by equation (8) in terms of far-field parameters [19]:

$$\text{ECC}_F = \frac{\left|\iint_{4\pi} [\vec{A}_i(\theta,\phi) * \vec{A}_j(\theta,\phi)] d\omega\right|^2}{\iint_{4\pi} \left|\vec{A}_i(\theta,\phi)\right|^2 \iint_{4\pi} \left|\vec{A}_j(\theta,\phi)\right|^2} \qquad (8)$$

where, $\vec{A}_i(\theta,\varphi)$ and $\vec{A}_j(\theta,\varphi)$ presents the 3-D far-field pattern when port-i and port-j is excited respectively. In the case of MIMO antenna design, ECC reduction with the help of far-field parameters is highly desirable.

Diversity gain (DG) is another important parameter. It tells us the information about the SNR at the receiver side or the effectiveness of diversity. Mathematically, DG is directly related to ECC as follows [19]:

$$DG = 10\sqrt{1 - (ECC)^2} \tag{9}$$

For an efficient antenna design, the value of ECC must be less than 0.5 while, the diversity gain is approx. 10 dB within the operating frequency band [19].

Mean effective gain (MEG) is defined as the MIMO antenna gain when some predefined wireless environmental conditions are given. Mathematically, it can be calculated as follows [19]:

$$MEG_1 = 0.5[1 - |S_{11}|^2 - |S_{12}|^2] \tag{10a}$$
$$MEG_2 = 0.5[1 - |S_{12}|^2 - |S_{22}|^2] \tag{10b}$$

where MEG_1 and MEG_2 are defined as the mean effective gain at port-1 and port-2 respectively. For better diversity performance, the difference between MEG_1 and MEG_2 should be less than 3-dB.

Channel Capacity Loss (CCL) measures the data rate up to which signal is moved uniformly over the communication system. It is measured by using the following formula [20]:

$$CCL = -\log_2 \det \begin{pmatrix} \beta_{11} & \beta_{12} \\ \beta_{21} & \beta_{22} \end{pmatrix} \tag{11}$$

where,
$$\beta_{11} = 1 - (|S_{11}|^2 + |S_{12}|^2)$$
$$\beta_{22} = 1 - (|S_{22}|^2 + |S_{21}|^2)$$
$$\beta_{12} = -(S_{11}^* S_{12} + S_{21}^* S_{12})$$
$$\beta_{21} = -(S_{22}^* S_{21} + S_{12}^* S_{21})$$

8.3 EMERGING APPLICATIONS OF PRINTED ANTENNAS

8.3.1 Antennas for 5G Mobile Communication

International Telecommunication Union (ITU) declaration of 5G frequency bands include 3.4–3.6 GHz, 5–6 GHz, 24.25–27.5 GHz, 37–40.5 GHz and 66–76 GHz bands and Federal Communications Commission (FCC) allocated spectrum lies in the frequency range of 27.5–28.35 GHz [1, 2]. The majority of countries are using the 26/28 GHz band for 5G / millimetre wave communications. The 5G mobile communication in the sub-6 GHz band is intended for providing faster, reliable multiple applications with improved system capacity which requires integration of multiple and diverse wireless services in a single mobile unit. These increasing demands significantly aim towards designing an antenna structure with the ability to switch/readjust its characteristics as per changing requirements like antenna capable of resonating at multiple frequency bands by using defected ground structure (DGS), split-ring resonators (SRR), supporting multiple higher-order modes, arrays, MIMO, and beamforming are some other commonly used techniques [2-6]. Another important aspect of 5G antenna design is the size reduction in view of keeping the device compact and portable which has forced researchers for inventing techniques such as shorting walls, inductive/ capacitive loading and using metamaterials for antenna size miniaturization.

Further, for reliable 5G communication at higher frequencies, it is required to establish point-to-point and point-to-multipoint wireless links in order to compensate for higher propagation losses. This demands higher antenna gain, quick beamforming with wider beam scanning. Hence, reconfigurable antennas are now extensively being used for covering diverse wireless applications operating in a wider frequency range as well as simultaneously delivering the same good performance as compared to using different, multiple antennas that too without increasing the size [21]. There are a variety of antenna reconfigurability techniques that ensure better signal reception, like; polarization reconfigurability accounts for reducing multipath fading and co-channel interference; frequency reconfigurability provides frequency tuning over diverse frequency bands and pattern reconfigurability supports automated beam steering in any desired direction. As shown in Figure 8.4 (a), a frequency reconfigurable antenna was discussed by H. Dildar et. al. having printed over

FR4 substrate employing pin diode to achieve the required frequen-cy reconfigurability with a single band at 3.5 GHz, dual bands at 2.6 GHz and 6.5 GHz or triple band (1.8, 4.8, and 6.4 GHz) thus covering sub-6 GHz 5G frequency spectrum (2.1, 2.6, 3.5, and 4.8 GHz) along with other applications of GSM, UMTS, 4G-LTE, WiMAX, WLAN Wire-less networks and providing efficiency 84 % and gain 1.2 to 3.6 dBi [22]. X. Zhou et. al. demonstrated a CPW feed graphene printed green MIMO antenna which was low-cost, conformal, flexible and environmental-friendly [23]. The proposed design exhibited in Fig-ure 8.4(b) achieved fractional bandwidth of 53.71% (2.22 GHz to 3.85 GHz) and was covering diverse frequency bands of mobile communication (LTE, 4G, 5G, sub 6 GHz), WiMAX (2.5 and 3.5 GHz), and WLAN (2.4 and 3.6 GHz) with ECC below 0.2×10^{-6}.

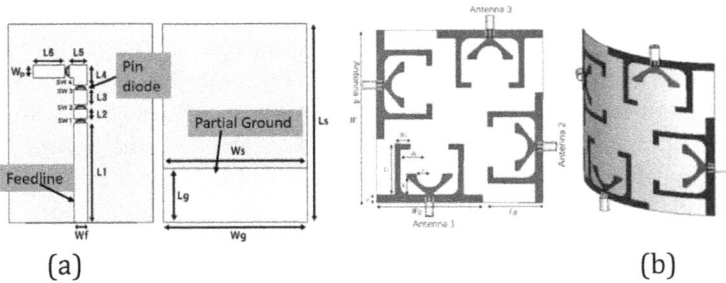

Figure 8.4 Antennas for 5G communication (a) Frequency reconfigurable antenna [22], (b) flexible graphene MIMO antenna [23]. Reproduced with permission.

Due to the already existing wireless applications at the sub-6 GHz frequency range, the available channel capacity and bandwidth are very restricted which has impelled researchers to move towards the mm-wave band for 5G communication. Another paper by S. Tariq et. al. [11] successfully showed a MIMO antenna design (Figure 8.5) with four elements (multiple-input multiple-output) and a super-strate layer of circular split ring meta resonator (CSRR) array in fre-quency range 24.5 to 26.5 GHz where integration of meta-surface provided the gain enhancement up to 10.27 dBi.

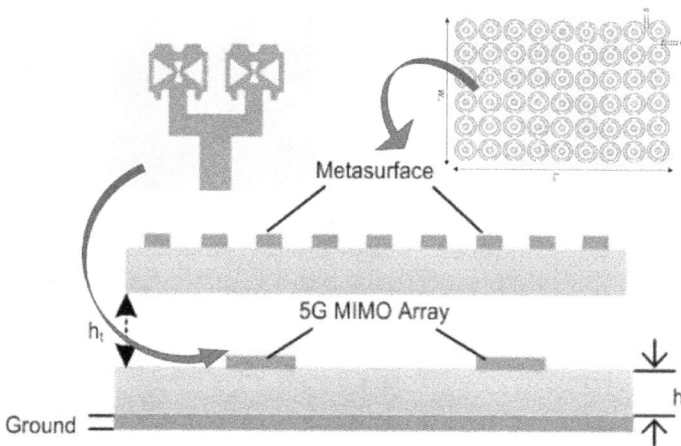

Figure 8.5 Metasurface superstrate integrated 5G MIMO antenna array [11]. Reproduced with permission.

8.3.2 Antenna for IoT Applications

Antennas for IoT applications generally require some important things such as high data rate, and compactness. In order to achieve a high data rate, there are two ways: wider impedance bandwidth or improved signal to noise ratio. Generally, the method of improved bandwidth is utilized for short-distance communication such as RFID. Otherwise, this method creates a large interference with the adjacent channels and enhances the noise level. MIMO antenna technique is the other method to enhance the data rate without changing the signal power level. Filtenna (i.e., filter+antenna) is widely used in IoT applications. These antennas have both filtering as well as radiation ability. Pan et.al. proposed a dielectric resonator-based filtenna [24]. In the given antenna, authors have excited the rectangular ceramic material with the assistance of a conformal strip along with a dual unequal horizontal microstrip stub. For the proper explanation, the authors have proposed four designed steps:

Antenna-1: Rectangular ceramic excited with conformal strip
Antenna-2: Rectangular ceramic excited with conformal strip and small microstrip stub
Antenna-3: Rectangular ceramic excited with conformal strip and large microstrip stub

Antenna-4: Rectangular ceramic excited with conformal strip and the combination of small and large microstrip stub

All these design steps are shown in Figure 8.6. Its corresponding $|S_{11}|$ and gain curves are shown in Figure 8.7. From the gain curve, it can be observed that bandpass filter response is achieved with the combination of both conformal strips as well as long and short microstrip stubs. The combination of a short microstrip stub and the conformal strip is accountable for band rejection at the higher frequency side. On the other hand, a combination of a long microstrip stub and the conformal strip is accountable for band rejection at the lower frequency side.

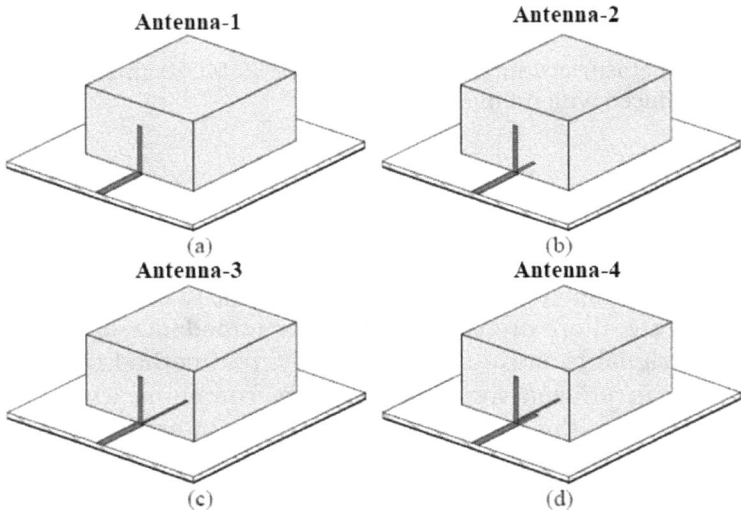

Figure 8.6 Step by Step design process for filtenna [24]. Reproduced with permission.

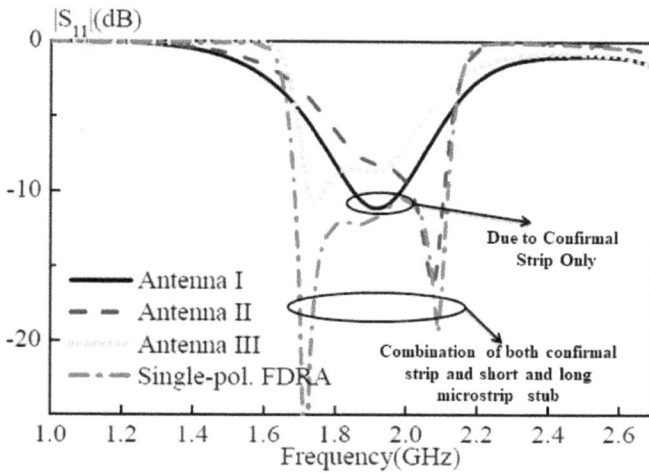

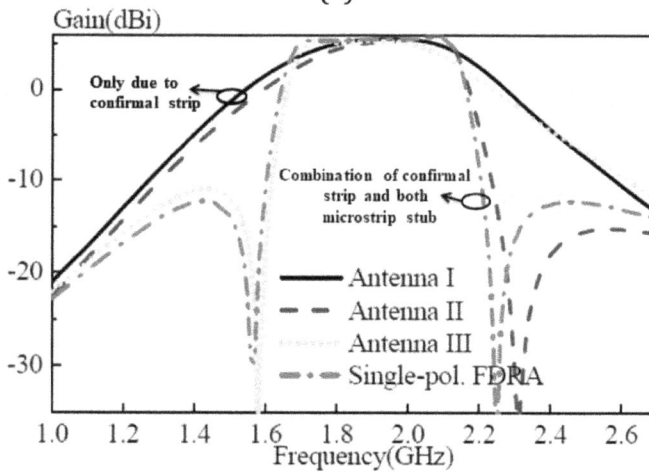

Figure 8.7 |S₁₁| and Gain variation of different antenna design steps [24]. Reproduced with permission.

In order to understand the reason, E-field is plotted at the operating frequency as well as at the frequency point where the gain value decreases suddenly as shown in Figure 8.8. From the E-field plot, it is clear that the operating band is due to $TE_{1\delta 1}$ mode. The radiation nulls are generated at 1.56 GHz and 2.25 GHz because two 180° out of phase-field lines cancel out each other.

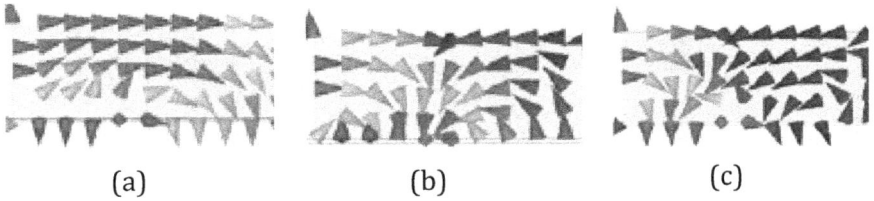

(a)　　　　　　　　　　　(b)　　　　　　　　　　　(c)

Figure 8.8 E-field variation (a) 1.87 GHz (b) 1.56 GHz (c) 2.25 GHz [24]. Reproduced with permission.

Jin et.al. proposed a coaxial probe fed patch antenna with filtering characteristics [25]. The authors have incorporated the pair of dual slits in the patches for getting filtering features. Figure 8.9 shows its geometrical layout. Figure 8.10 shows $|S_{11}|$ and Gain variation of proposed patch-based filtering antenna. From Figure 8.10, it can be said that pair of slits with the lesser gap is accountable for radiation null at a higher frequency while slits with a larger gap are accountable for null at a lower frequency. In order to understand the basic phenomenon, the current distribution at 5.13 GHz and 4.43 GHz is shown in Figure 8.11. Dual slits create 180^0 out of phase in the current distribution at 4.43 GHz and 5.95 GHz. It will create radiation null at these frequencies.

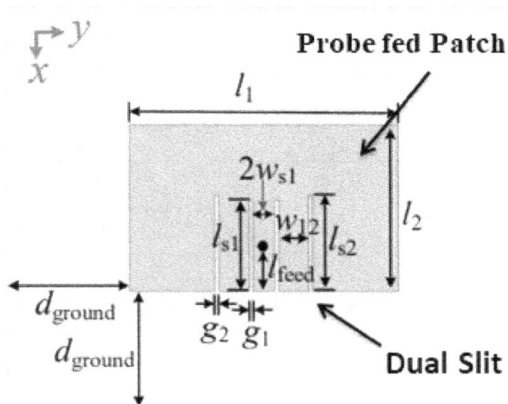

Figure 8.9 Patch based filtering antenna [25]. Reproduced with permission.

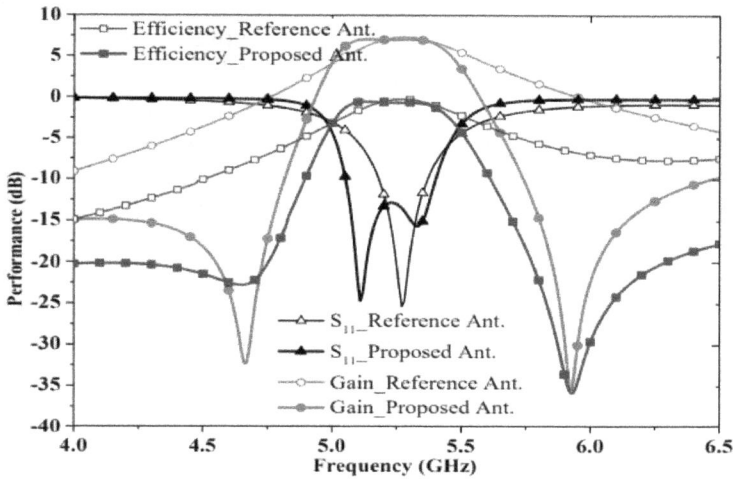

Figure 8.10 |S₁₁| and Gain variation of patch based filtering antenna [25].
Reproduced with permission.

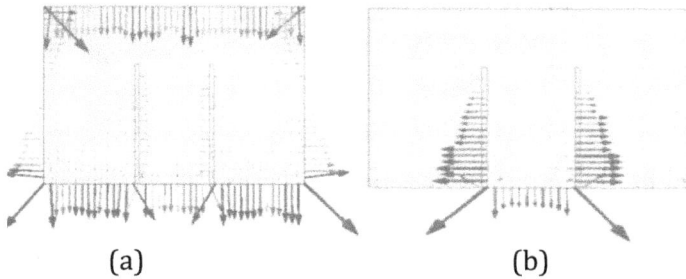

Figure 8.11 Current Distribution on patch at (a) 5.13 GHz (b) 4.43 GHz
[25]. Reproduced with permission.

8.3.3 Antennas for Satellite Communication

Antenna design for satellite communication requires high gain and
circular polarization characteristics. There are lots of techniques
available in the literature for enhancing the gain such as the use of
superstrate [26], the concept of array [27] and the use of frequency-
selective surfaces [28]. The mandatory requirement of antennas for
satellite communication is the creation of circular polarization (CP)
characteristics. In order to create CP waves, two necessary condi-

tions must be fulfilled: (a) creation of degenerated orthogonal modes; and (b) 90⁰ phase shift between the orthogonal modes [14]. Kandasamy et.al. presented a dual-band slot based CP antenna [29]. In the proposed antenna design, a split ring is placed on the bottom part of the substrate while the top truncated slot is etched. Its geometrical layout, $|S_{11}|$ characteristics, antenna gain and axial ratio variation is shown in Figure 8.12. Dual connected split-ring resonator and perturbed aperture is accountable for CP wave at upper and lower frequency band respectively [29]. The main drawback of this antenna structure is high back radiation. It results in a lower gain value. However, the authors improve the gain value by using the metallic cavity on the backside of the aperture.

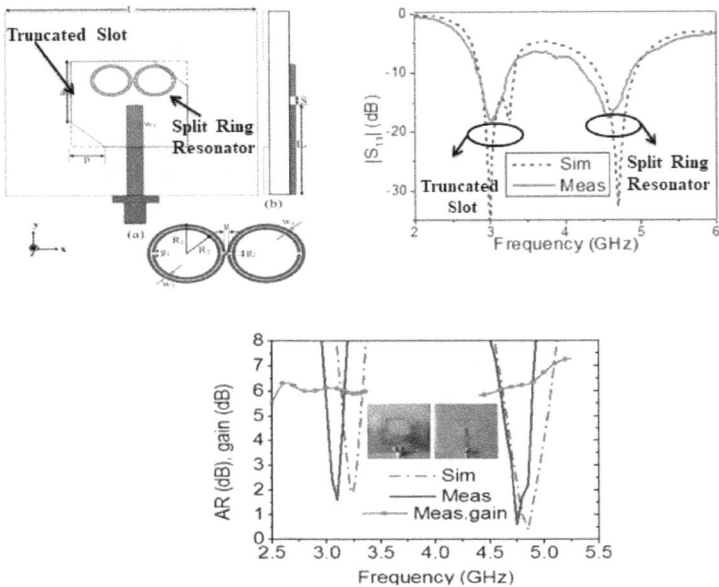

Figure 8.12 CP Antenna Geometry, its $|S_{11}|$, gain and axial ratio [29]. Reproduced with permission.

8.3.4 Antennas for Body Area Network (BAN) Communication

Body area networks (BAN) has grown substantial attraction due to their wide range of applications, such as real-time safety screening, sports, medical emergency response, fitness tracking, care of elderly, children and underprivileged persons. A wearable antenna is an indispensable component in any BAN system which allows a person to remain under real-time communication and monitoring over the network. Ultra-wideband (UWB) antenna due to its large bandwidth of 7.5 GHz (3.1 to 10.6 GHz) is the natural choice in order to reduce the effective power emission < -43dBm/ MHz, which provides longer battery life for wearable/ implantable medical use, reduced electromagnetic sensitivity for continual on-body operation, less interference with surroundings and higher supported data rate (order of Mbps) [21]. Practical design aspects of wearable antennas are coupling between an antenna and human body, robustness in terms of antenna deformations due to body movement, varying temperature/humidity, wash resistance, compact size, low cost, easy fabrication, wideband/ multiband and high mechanical strength. Plenty of work has been done and currently going on towards the design of flexible and resistant wearable antennas. Different reviews of the state of artwork for the on-body wearable antenna are available [7, 9]. The critical antenna design challenges are minimal coupling to the human body and to withstand the characteristics under different deformations. Specific absorption rate (SAR) is a key parameter that signifies electromagnetic power absorption by the human body when an RF transceiver is placed in close proximity or implanted. As per FCC norms, the standard SAR value is around 1.6 W/kg in 1 g of tissue, and it is defined as [13].

$$SAR = \frac{\sigma|E^2|}{\rho} \qquad\qquad (12)$$

where, ρ is tissue mass density, σ is its conductivity, and E is the applied electric field.

Lin *et al.* [8] have proposed an on-body ultra-wideband complete textile antenna architecture made up of lightweight fabric (polyester) and taffeta (copper) for early diagnostics of pathogens interior to the human body as shown in Figure 8.13. The textile antenna provided a compact and fully flexible UWB bandwidth of 109% (1.2 to 4.1 GHz) and a gain of around 2.9dBi. The on-phantom simulation results showed that in the close vicinity of the body also antenna performance does not affect.

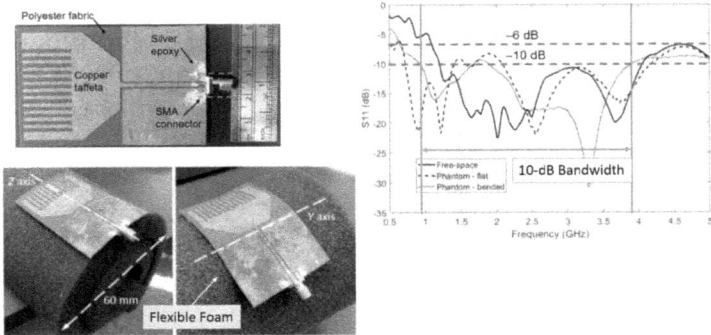

Figure 8.13 CPW fed flexible textile antenna for body area communication (a) complete antenna structure (b)antenna on flexible foam (c) ultra-wide bandwidth in return loss plot [8]. Reproduced with permission.

A metamaterial-based wearable antenna has the added advantages of miniaturization, multifunction, multi-band and broadband frequency. A pattern-reconfigurable, flexible metamaterial wearable folded slot antenna is shown in Figure 8.14 where reconfigurability was achieved using pin diode and covering WBAN & WiMAX bands [21]. The artificial magnetic conductor (AMC) surface is placed as a substrate over the antenna in order to enhance the pattern and reduce SAR. When diode is on, this antenna exhibits single-band 2.45 GHz (2.4–2.5 GHz) for WBAN applications and when p-i-n diode is kept OFF, it will work as a dual-band with resonant frequencies as 2.45 GHz (2.35–2.52 GHz) and 3.3 GHz (3.28–3.38 GHz) for WiMAX wireless applications. The observed maximum SAR value for one gram with AMC surface through simulation is reduced to 0.29 W/kg as compared to 2 W/kg without AMC surface which lies within permissible limits.

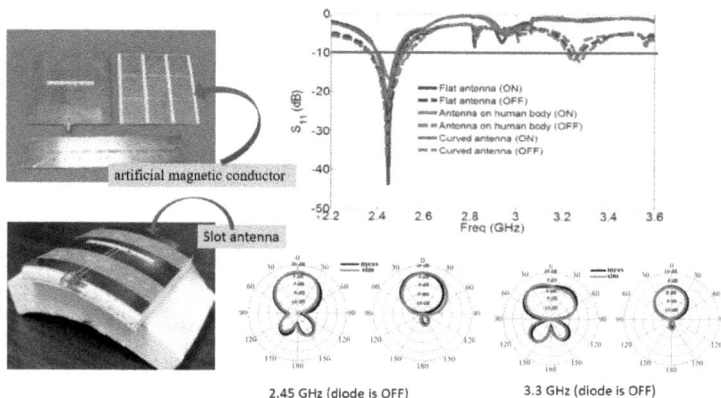

2.45 GHz (diode is OFF)   3.3 GHz (diode is OFF)

Figure 8.14 Wearable, flexible folded slot antenna, reflection plot and directional radiation pattern at two different wireless communication frequencies (2.45 & 3.3 GHz) [21]. Reproduced with permission.

8.3.5 Antennas for Low Range radar Communication

The advancements in high gain, compact and printed antenna versions have allowed the adaptation of heavy and large sized radar systems to a handheld, easily portable, non-contact communication device for real-time monitoring and early warning system design, for example, automotive radar sensors in driving assistance, unmanned aerial vehicles, remote monitoring of vital signs (heartbeat and respiration), automated fruits and goods quality monitoring and segregation are some trending and demanding application areas.

8.3.6 Vehicular Communication

A smart, collision-avoidance feature has become an essential attribute for today's automatic and safe driving which provides information such as approaching vehicles, road environments with respect to distance, speed, and angle (cross-range) [30]. Automotive antennas technologist faces many challenging tasks such as proposed structure be fit for mass production, able to function in an extreme temperature range of 40 °C to 85 °C and even up to 105 °C, be shockproof, and compact in size. For short-range communication, 24.1– 24.3 GHz frequency range in the ISM band is used but in future practically all long/ medium range antennas will function in the 76–81GHz range. Range resolution (ΔR) i.e., separation between two targets in an automotive radar sensor is given by equation (13) [15]

$$\Delta R \geq \frac{c_o}{2\Delta f} \tag{13}$$

$$\phi = 60° \frac{\lambda}{D} \tag{14}$$

with c_o is the speed of light and Δf is instantaneous modulation bandwidth. In early automotive radar systems, the typical value of range was around one metre while for current short-range systems it is few centimetres due to broader bandwidth of up to 5 GHz. The cross-range (angular) resolution of the sensor is determined by the antenna 3 dB beamwidth (in degrees) given by equation (14). Here, λ is the wavelength, and D is the maximum antenna dimension. For far range automotive applications, in general, the antenna beam-width is around 3 ~4 degrees, antenna diameter from 60 to 100 mm with a required gain of approximately 30 dB. In order to properly illuminate the scene of interest, beam tilting or electronic beam scanning may be achieved using either a phased array antenna or digital beamforming. A new cognitive receiver for automotive imaging radars was evaluated in a single and multi-target scenario utilizing Bobrovsky-Zakai bound (BZB) optimization criterion for achieving high angular resolution with additional immunity towards ambiguity for cognitive antenna selection (CASE) as shown in Figure 8.15 [31].

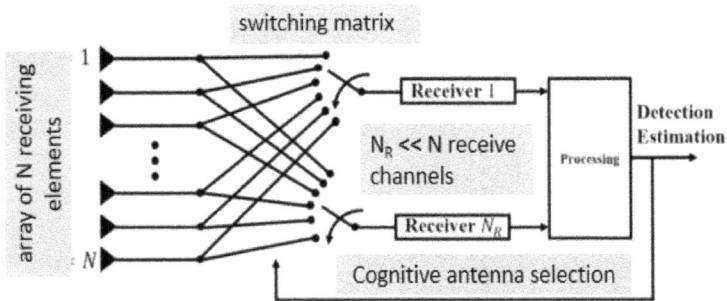

Figure 8.15 Demonstration of cognitive antenna selection in a multi-target vehicular communication using automotive collision avoidance radar [31]. Reproduced with permission.

W. Wang et. al. has designed a wide band, compact MIMO Vivaldi antenna for intelligent vehicular communication using four modified Vivaldi antenna placed orthogonally coupled with metasurface and parasitic stub for improved gain and wider bandwidth as shown in

Figure 8.16 [32]. The fractional bandwidth covered was from 7.55 GHz to 22.85 GHz with high isolation between ports of around 18 dB. The measured gain ranges between 6.9 to 11.5 dBi over this full frequency range. Further, a detailed study was also done in simulation in different practice environments while mounting MIMO antenna assembly over car roof as shown in Figure 8.16(c) in the hot-humid, cold and dry weather conditions. A very good gain and radiation pattern were observed with a good acceptance level for the vehicle to vehicle (V2V) and vehicle to device (V2D) communication.

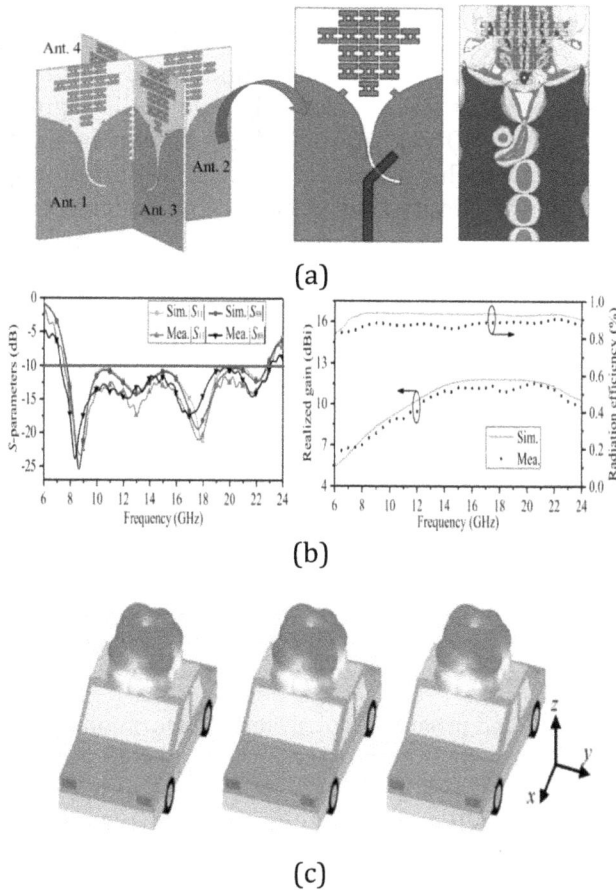

(a)

(b)

(c)

Figure 8.16 (a) MIMO Vivaldi antenna with improved field distribution (b) reflection coefficient, gain and radiation efficiency (c) antenna mounted over vehicle simulation depicting radiation pattern in hot-humid, cold and dry environments [32]. Reproduced with permission.

8.3.7 Remote Healthcare Monitoring Systems

Electromagnetics-based remote health monitoring and early warning systems are very optimistic alternatives for continuous heartbeat and respiration measurements of infants, sleep apnoea, elderly patients and for round, the clock centralized observations in hospitals. These noncontact doppler based radar systems offer lesser design and fabrication cost, excellent performance, and have the unique advantage of exposure through opaque objects like walls and clothing. A short-range vital sign monitoring doppler CW radar has been proposed by [33] using concurrent dual-beam phased-array antenna operating at 2.4 GHz Using the MIMO beamforming technique. Another, 24 GHz frequency modulated continuous wave (FMCW) doppler radar was successfully able to distinguish between two humans sitting 40 cm apart [10]. As shown in Figure 8.17, a higher resolution much needed for discrimination among multiple subjects in close proximity was achieved by designing a frequency scanning fan beam array antenna having broad impedance bandwidth ranging from 2.8 GHz to 6 GHz integrated with BPF [34].

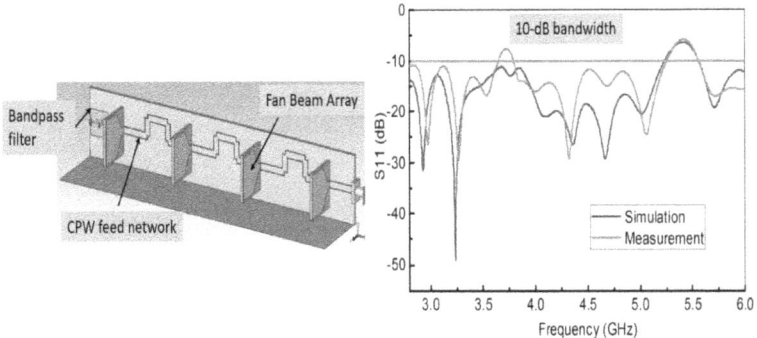

Figure 8.17 A CPW fed broadband array antenna for multiple human discrimination in close proximity [34]. Reproduced with permission.

In Figure 8.18, a UWB radar system employing a circularly polarized, elliptical-shaped antipodal vivaldi antenna array has been shown by K. K. M. Chan et.al. [35]. Here, circular polarization has been used to improve the RCS of weak received vital signals and to provide stability and robustness towards any fluctuations due to human body movement. Here, wide bandwidth and good axial ratio achieved over 3 to 10 GHz frequency have enhanced the sensitivity and accuracy of the retrieved vital sign of breathing and heart rate as shown by Figure 8.18(b).

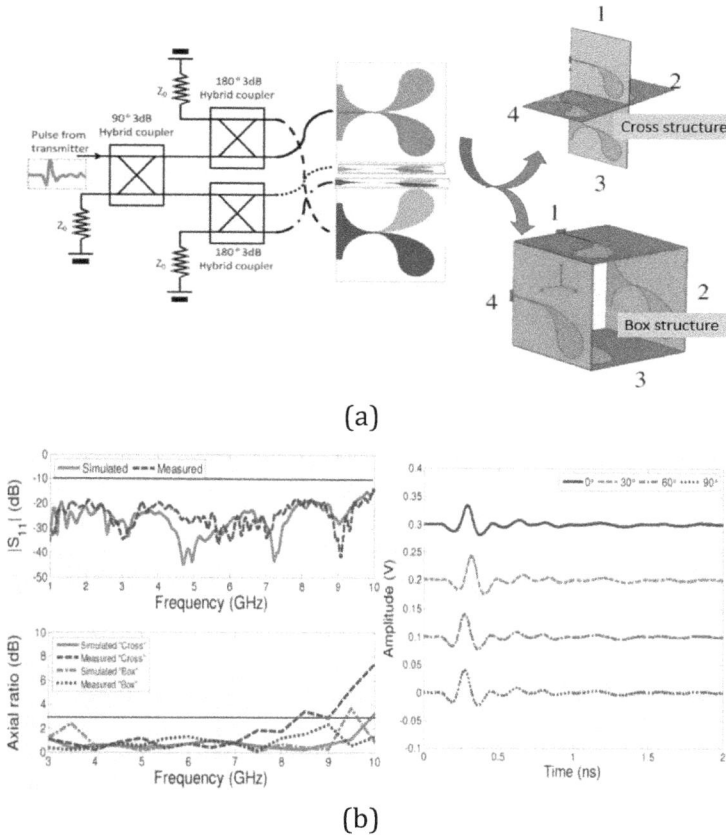

(a)

(b)

Figure 8.18 (a) UWB radar system utilizing assembly of four circularly polarized vivaldi antenna, (b) reflection coefficient, axial ratio and received time-domain signal plot [35]. Reproduced with permission.

8.4 CONCLUSION

A comprehensive review of current antenna technology in wireless communication has been discussed. Firstly, important antenna attributes of the radiation pattern, gain, bandwidth, diversity parameters etc. were explained with respect to optimizing antenna performance. A diverse flavour of antenna designs present in literature are characterized by their respective features, benefits, and limitations and are chosen on the basis of particular application requirements. Salient findings of this chapter are as follows:

- Microstrip printed antenna technology however suffers from material losses and performance limitations at higher frequencies but affords valuable features of lesser cost, conformability and ease of integration for compact, smart RF system design hence it is the most viable and researched choice in wireless communication.

- 5G mobile communication, IoT and satellite communication applications due to the inclusion of multiple applications require multiband/ broadband antenna along with reconfigurability to enable quick switching between different bands, smaller antenna size to provide a compact mobile device, larger bandwidth to support higher data transfer rate and improved signal to noise ratio.

- Vehicular communication demands rapid beam steering and wider bandwidth for improved system cross-range and downrange resolution, respectively. Further, real-time collision avoidance and anti-accident alarming system for road safety require a high information transfer rate supported by a wideband antenna.

- Body area communication applications require a flexible, conformal antenna design, as well as minimum SAR value, is needed to reduce electromagnetic wave absorption to the human tissue. Further, losses due to flexible substrate and bending of woven fabric antenna at certain angles deviate antenna performance which is a challenging task to address for these researchers.

- The other key research gaps needed to be filled are higher propagation / material losses, electromagnetic interference due to close proximity of antenna elements, integration of a greater number of features in a minimum possible space due to the apparition of novel wireless communication applications.

In near future with the advent of more robust antenna system designs interesting features and opportunities for communication will be offered to the end-user. The benefitted features of these well-known printed antennas are also being explored in emerging fields of RFID, gas sensing, bio-sensing, nano-structure deposition, energy harvesting etc., and one should be completely cheerful about the advent of new paradigms based on these antennas.

References

1. Federico, G., Caratelli, D., Theis, G., and Smolders, A. B. (2021) A Review of Antenna Array Technologies for Point-to-Point and Point-to-Multipoint Wireless Communications at Millimeter-Wave Frequencies. *International Journal of Antennas and Propagation.*
2. Jensen, M. A., and Wallace, J. W. (2004) A review of antennas and propagation for MIMO wireless communications. *IEEE Transactions on antennas and propagation, 52*(11), 2810-2824.
3. Patel, R., Desai, A., Upadhyaya, T., Nguyen, T. K., Kaushal, H., and Dhasarathan, V. (2021) Meandered low profile multiband antenna for wireless communication applications. *Wireless Networks, 27*(1), 1-12.
4. Agarwal, S. (2021) Concurrent 60/94 GHz SIR Based Planar Antenna for 5G/MM-Wave Imaging Applications. *Wireless Personal Communication.*
5. Agarwal, S. (2020) Design of on-Chip Compatible Concurrent Dual Band Millimeter Wave Antenna. *Progress In Electromagnetics Research C*, 102, 213-223.
6. Anguera, J., Andújar, A., Huynh, M. C., Orlenius, C., Picher, C., and Puente, C. (2013) Advances in antenna technology for wireless handheld devices. *International Journal of Antennas and Propagation.*
7. Mahmood, S. N., Ishak, A. J., Ismail, A., Soh, A. C., Zakaria, Z., and Alani, S. (2020) ON-OFF Body Ultra-wideband (UWB) Antenna for Wireless Body Area Networks (WBAN): A Review. *IEEE Access, 8*, 150844-150863.
8. Lin, X., Chen, Y., Gong, Z., Seet, B. C., Huang, L., and Lu, Y. (2020) Ultrawideband textile antenna for wearable microwave medical imaging applications. *IEEE Transactions on Antennas and Propagation, 68*(6), 4238-4249.
9. Yan, S., Soh, P. J., and Vandenbosch, G. A. (2018) Wearable ultrawideband technology—A review of ultrawideband antennas, propagation channels, and applications in wireless body area networks. *IEEE Access, 6*, 42177-42185.
10. Lee, H., Kim, B. H., Park, J. K., Kim, S. W., and Yook, J. G. (2019) A resolution enhancement technique for remote monitoring of the vital signs of multiple subjects using a 24 GHz bandwidth-limited FMCW radar. *IEEE Access, 8*, 1240-1248.
11. Tariq, S., Naqvi, S. I., Hussain, N., and Amin, Y. (2021) A Metasurface-Based MIMO Antenna for 5G Millimeter-Wave

Applications. *IEEE Access*, 9, 51805-51817.

12. Agarwal, S., Pathak, N. P., and Singh, D. (2013) Concurrent 85GHz/94GHz slotted gap coupled parasitic microstrip antenna for millimeter wave applications. *National Conference on Communications (NCC)*, India, pp. 1-5.

13. IEEE Recommended Practice for Determining the Peak Spatial-Average Specific Absorption Rate (SAR) in the Human Head From Wireless Communications Devices: Measurement Techniques—Redline (2013). *IEEE Std 1528- 2013 (Revision of IEEE Std 1528-2003)—Redline*,1–500.

14. Balanis, C. A. (2016) *Antenna theory: analysis and design*, 4th edition, John Wiley & Sons, New Jersey.

15. Agarwal, S., Bisht, A. S., Singh, D., and Pathak, N. P. (2014) A novel neural network based image reconstruction model with scale and rotation invariance for target identification and classification for Active millimetre wave imaging. *Journal of Infrared, Millimeter and Terahertz Waves*, 35(12), 1045-1067.

16. Agarwal, S., Singh, D. (2016) Optimal Non-Invasive Fault Classification Model for Packaged Ceramic Tile Quality Monitoring Using MMW Imaging. *Journal of Infrared, Millimeter and Terahertz Waves*, 37(4), 394–413.

17. Yaghjian, A. D., and Best, S. R. (2005) Impedance, bandwidth, and Q of antennas. *IEEE Transactions on Antennas and Propagation*, 53(4), 1298-1324.

18. Pozar, D. M. (2011) *Microwave engineering*, 4th edition, John Wiley & Sons, New Jersey.

19. Sharawi, M. S. (2014) *Printed MIMO antenna engineering*, Artech House.

20. Sharma, A., Das, G. and Gangwar, R. K. (2018) Design and analysis of tri-band dual-port dielectric resonator based hybrid antenna for WLAN/WiMAX applications. *IET Microwaves, Antennas Propagation,* 12(6),986-992.

21. Saeed, S. M., Balanis, C. A., Birtcher, C. R., Durgun, A. C., and Shaman, H. N. (2017) Wearable flexible reconfigurable antenna integrated with artificial magnetic conductor. *IEEE Antennas and Wireless Propagation Letters*, 16, 2396-2399.

22. Dildar, H., Althobiani, F., Ahmad, I., Khan, W.U.R., Ullah, S., Mufti, N., Ullah, S., Muhammad, F., Irfan, M. and Glowacz, A., (2021) Design and Experimental Analysis of Multiband Frequency Reconfigurable Antenna for 5G and Sub-6 GHz Wireless Communication. *Micromachines*, 12(1), 32.

23. Zhou, X., Leng, T., Pan, K., Abdalla, M. A., Novoselov, K. S., and Hu, Z. (2021) Conformal screen printed graphene 4× 4 wideband MIMO antenna on flexible substrate for 5G communication and IoT applications. *2D Materials*, 8, 045021.

24. Pan, Y. M., Hu, P. F., Leung, K. W. and Zhang, X. Y. (2018) Compact single-/dual-polarized filtering dielectric resonator antennas. *IEEE Transactions on Antennas Propagation,*66(9), 4474-4484.

25. Jin, J. Y., Liao, S. and Xue, Q. (2018) Design of filtering-radiating patch antennas with tunable radiation nulls for high selectivity. *IEEE Transactions on Antennas Propagation,* 66 (4), 2125-2130.

26. Lee, R. and Lee, K. (1988) Gain enhancement of microstrip antennas with overlaying parasitic directors. *Electronics Letters,* 24(11), 656-658.

27. Chen, H.-D., Wu, J.-Y. and Chiu, T. W. (2012) Broadband high-gain microstrip array antennas for WiMAX base station. *IEEE transactions on antennas propagation,* 60 (8),3977-3980.

28. Sah, S., Mittal, A. and Tripathy, M. R. (2019) High gain dual band slot antenna loaded with frequency selective surface for WLAN/fixed wireless communication. *Microwave Optical Technology Letters,* 61(2), 519-525.

29. Kandasamy, K., Majumder, B., Mukherjee, J. and Ray, K. P. (2016) Dual-band circularly polarized split ring resonators loaded square slot antenna. *IEEE Transactions on antennas Propagation,* 64(8), 3640-3645.

30. Menzel, W., and Moebius, A. (2012) Antenna concepts for millimeter-wave automotive radar sensors. *Proceedings of the IEEE,* 100(7), 2372-2379.

31. Tabrikian, J., Isaacs, O., and Bilik, I. (2021) Cognitive Antenna Selection for Automotive Radar Using Bobrovsky-Zakai Bound. *IEEE Journal of Selected Topics in Signal Processing,* 15(4), 892-903.

32. Wang, W. and Zheng, Y. (2021) Wideband Gain Enhancement of a Dual-Polarized MIMO Vehicular Antenna. *IEEE Transactions on Vehicular Technology,*70(8), 7897- 7907.

33. Nosrati, M., Shahsavari, S., Lee, S., Wang, H., and Tavassolian, N. (2019) A concurrent dual-beam phased-array Doppler radar using MIMO beamforming techniques for short-range vital-signs monitoring. *IEEE Transactions on Antennas and Propagation,* 67(4), 2390-2404.

34. Rahman, M., NaghshvarianJahromi, M., Mirjavadi, S. S., and Hamouda, A. M. (2018) Bandwidth enhancement and frequency scanning array antenna using novel UWB filter integration technique for OFDM UWB radar applications in wireless vital signs monitoring. *Sensors,* 18(9), 3155.

35. Chan, K. K., Tan, A. E., and K. Rambabu (2013) Circularly Polarized Ultra-Wideband Radar System for Vital Signs Monitoring. *IEEE Transactions on Microwave Theory and Techniques,* 61(5), 2069-2075.